U0155099

管派古琴制作与修复

Making and Restoration of Guan-School Guqin

上册
Volume 1

昭闻 著

by Zhao Wen

Convention for the Safeguarding of the Intangible Cultural Heritage

The Intergovernmental Committee for the Safeguarding of the Intangible Cultural Heritage has incorporated

The Guqin and its music

in the Representative List of the Intangible Cultural Heritage of Humanity upon the proposal of China

Incorporation in this List contributes to ensuring better visibility of the intangible cultural heritage and awareness of its significance, and to encouraging dialogue which respects cultural diversity

Date
4 November 2008

Director-General of UNESCO

陕西省哲学社会科学重大理论与现实问题研究项目
Research Project on Major Theoretical and Practical Issues in Philosophy and Social Sciences in Shaanxi Province
程刚省级技能大师工作室学术成果
Academic Achievements of Cheng Gang Skill Master Studio in Shaanxi Province

中国书店

图书在版编目（CIP）数据

管派古琴制作与修复：汉、英 / 昭闻著 . — 北京：中国书店，2022.9

ISBN 978-7-5149-2840-2

Ⅰ.①管… Ⅱ.①昭… Ⅲ.①古琴－乐器制造－汉、英②古琴－修复－汉、英 Ⅳ.① TS953.24

中国版本图书馆 CIP 数据核字 (2021) 第 178259 号

管派古琴制作与修复

昭闻 著

责任编辑：姚文杰

出版发行： 中国书店

地　　址：北京市西城区琉璃厂东街 115 号

邮　　编：100050

印　　刷：陕西龙山海天艺术印务有限公司

开　　本：889mm×1194mm　16 开

版　　次：2022 年 9 月第 1 版第 1 次印刷

印　　张：26.5

字　　数：150 千字

印　　数：1-3000

书　　号：ISBN 978-7-5149-2840-2

定　　价：520.00 元

| 前 言 |

20 世纪 50 年代，我是北京民族乐器厂的技术研发人员，那时候在厂里的工作主要是研发和制作古筝、琵琶、二胡、阮等民族乐器，居住的地方与管平湖先生相邻，所以比较熟悉。

1958 年夏天，吴景略和管平湖先生找到我，希望我制作一些新的古琴，那时候历史老琴特别多，除了管先生和他老师杨时百老先生以外，基本没有再做新琴的人了。管平湖先生挑了一张音色、手感、外观自己都很满意的清代仲尼式琴作为参考，印象中那张琴比传统老琴的琴面要平一些，弧度没有那么大。我动手测量、画图、制作，管先生在一旁看着，他有时候也会亲自上手制作一会儿，特别是他认为比较重要的工序时，会一边动手一边讲解，偶尔还会用他那一双有力的大手对琴材进行敲击，并用耳朵贴得很近地去听，听完后若有所思地停顿一会儿，然后再让我也听听。

当时一共制作了三张琴，两张仲尼式、一张伏羲式，合琴以后由于古琴相对其他乐器使用生漆量很大，我们害怕导致乐器厂其他同事过敏，所以就把髹漆工作换到了我居住的地方，管先生每天下班以后到我家看着我工作，吴景略先生偶尔也会来小坐一会儿。

1959 年秋天，三张琴制作好以后，北京民族乐器厂发帖邀请北京的古琴界音乐家，在东琉璃厂文化馆召开了新琴鉴定会，到场的嘉宾有查阜西、溥雪斋、吴景略、程午嘉、王迪、许建等三十多人，那时候李祥霆还是学生，跟着老先生们帮忙干一些搬桌子、换琴等力所能及的事情。溥雪斋先生第一个上台试弹，他那时候已经有些耳背了，但觉得琴音很大，音色很好，就迟迟不愿意起来，弹了很长时间。现场音乐家们经过试弹和试听，最后一致给予其好评。鉴定会后由厂长亲自执笔撰写了新闻稿，刊登在了当天的《北京晚报》上。后来在吴景略先生的帮助下，民族乐器厂将这三张琴都卖到了国外，其中一张出现煞音、变形，还是由吴景略先生修缮的。

　　斫琴是一项对斫琴师要求很高的综合性技艺，我很感念我的老师管平湖先生，他那种对古琴事业执着和无私奉献的精神，深深影响了我的一生。我愿意将我毕生所学毫无保留地传授给每一位前来学习的学生。

　　谨以此书回馈古琴界。

二〇二〇年秋月
于北京虞田琴斋

田双坤

Preface

I served as a technical researcher at the Beijing National Musical Instrument Factory in the 1950s. My position involved developing and making national musical instruments such as Guzheng, Pipa, Erhu and Ruan. As I lived next to Mr. Guan Pinghu, I knew him very well.

In the summer of 1958, Mr. Wu Jinglue and Mr. Guan Pinghu came to me and hoped that I would make some new Guqin. At that time, there were many old Guqin. Yet apart from Mr. Guan and his teacher Mr. Yang Shibai, there were basically no master makers who made new qin. Mr. Guan Pinghu selected a Qing Dynasty Zhong Ni-style qin that he was satisfied with in timbre, hand feeling and appearance as a reference, which in my impression, was flatter and less radian than traditional ones. I then started measuring and drawing to make the qin. Mr. Guan watched me from the sidelines, and sometimes he would make it himself for a while. When it comes to the steps which he considered important, he would explain to me with his hands moving constantly. Occasionally he would use his strong big hands to tap the qin materials and stick his ears very close to them to listen. After listening, he would pause thoughtfully for a while, and then asked me to listen.

A total of three qins were made, two of Zhong Ni-style and one of Fuxi-style. Compared with other musical instruments, Guqin needs a large amount of raw lacquer in the lacquering process after assembling. To avoid causing allergies to my colleagues in the factory, we decided to lacquer it in my house. Mr. Guan would come to my house to watch me do my work every day after he got off work, and Mr. Wu Jinglue also occasionally came and stayed for a while.

In the autumn of 1959, shortly after the three qins were made, the Beijing National Musical Instrument Factory sent invitations to Beijing's Guqin musicians for a new qin appraisal session held at the East Liulichang Cultural Center, which was attended by a total of 30 musicians, including Zha Fuxi, Pu Xuezhai, Wu Jinglue, Cheng Wujia, Wang Di and Xu Jian. Li Xiangting, who was then a student, came to assist his teacher with some work that he was physically capable of doing, such as moving the tables and changing the qin. Mr. Pu Xuezhai was the first to try the newly made qins. Though hard of hearing, he found that the qin he played could make a huge sound and had a good timbre, so he was reluctant to get up and played for a long time. All the musicians present, after playing and auditioning, gave favorable comments on the three qins. After the appraisal session, the factory director personally wrote a press release and published it in the Beijing Evening News on that same day. Later, all the three qins were sold overseas by the National Musical Instrument Factory with the help of Mr. Wu Jinglue, who also repaired one of them that had problems with bad sound and deformation.

Guqin making is a comprehensive skill that demands high requirements for master makers. I am very grateful to my teacher Mr. Guan Pinghu, for his persistence and dedication to Guqin have swayed my life deeply. I myself am willing to pass on what I have learned to every student who comes to study with me without reservation. This book is, therefore, dedicated to everyone in the Guqin world.

Tian Shuangkun
Autumn 2020, at Yutian Guqin Studio, Beijing

目 录
Contents

人物简介

Profile

杨宗稷（1863—1932）

杨宗稷，字时百，号"九嶷山人"，师承于金陵琴派著名琴师黄勉之。湖南宁远县平田村（今湖南省永州市宁远县清水桥镇平田村）人，光绪二十七年（1901）赴京工作，先后任京师大学堂支应襄办、学部主事、邮电部侍郎、湖南南县知县等职，1917年辞官在京专事古琴，1921年应蔡元培之邀，任北京大学古琴导师。他是中国古琴重要门派"九嶷派"创始人，其传人有杨葆元（1899—1958）、彭祉卿（1891—1944）、关仲航（1896—1972）、管平湖（1897—1967）、李浴星（1908—1976）等。

杨宗稷不但是琴学的一代宗师，也是斫琴的一代大师，据其著作《琴学丛书》中《藏琴录》《琴余漫录》《琴学随笔》所记载，他自斫琴共计19床。他在京开设"九嶷琴社"之后，因和者众多，苦无多琴，遂派其仆人秦华在南县斫近一百四十琴，其音色与唐宋琴不相伯仲。但相对他所记载的斫琴数量而言，杨宗稷所修之琴更多，他在古琴声音品质方面有着独到的追求，其收藏的古琴，几乎每一床琴都经过破腹大修，究其原因，在于那个时代的历史老琴众多而易得。新制琴较少的原因在其《琴余漫录》卷一中也有说明："予深谙制琴法而未尝多制，以良材不易得。"

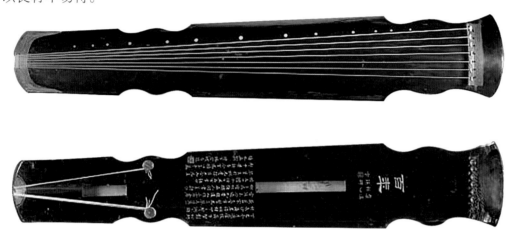

杨宗稷制宣和式"百年"琴　浙江省博物馆藏

Yang Zongji (1863—1932)

Courtesy name Shibai, art name "Jiuyi Shanren", studied under Huang Mianzhi, a great master of Jinling School of qin. A native of Pingtian Village, Ningyuan County, Hunan Province, he went to Peking to take up a post in 1901 (the 27th year of Guangxu's reign). There, he successively served as the deputy assistant and head of the faculty of the Imperial University of Peking, the assistant minister of the Ministry of Posts and Telecommunications, and the magistrate of Nan County, Hunan Province. He resigned in 1917 to devote himself completely to qin studies in Peking. In 1921, he was invited by Cai Yuanpei to serve as the Guqin mentor of Peking University. He is the founder of Jiuyi School, an important sect of Guqin in China. His disciples include Yang Baoyuan (1899-1958), Peng Zhiqing (1891-1944), Guan Zhonghang (1896-1972), Guan Pinghu (1897-1967), Li Yuxing (1908-1976), etc.

Aside from being a great master of Guqin studies, Yang Zongji is also a renowned master maker of Guqin. As recorded in "Record of Collected Qins", "Casual Extra Notes about Qin", and "Qin Study Jottings" in his Qinxue Congshu (Great Compendium of Qin Studies), he himself made a total of 19 qins, 9 of which were made by taking Muming-style qin for reference. After the "Jiuyi Guqin Society" was set up in Peking, due to the vast number of qin players and the lack of qin, Qin Hua, his servant, was sent to make nearly 140 qins in Nan County, with timbre comparable to that of qin made in Tang and Song Dynasties. The actual number of qin made by Yang Zongji is likely much higher than that is recorded. He has a unique pursuit in the sound quality of ancient qins. Almost every ancient qin of his collection has been overhauled. The reason lies in that plenty of aged qins are easily available while newly made ones are scarce, as stated in Volume 1 of "Casual Extra Notes about Qin": "The reason why I fail to make as many qins as possible is that good materials do not come easily."

管平湖（1897—1967）

　　管平湖，名平，字吉庵、仲康，号平湖，自称门外汉，江苏苏州（今江苏苏州）人。清代宫廷名画家管念慈之子，自幼酷爱艺术，弹琴学画皆得家传，并师从名画家金绍城，学花卉、人物，擅长工笔，笔法秀丽新颖，不为成法所拘，为湖社画会主要成员之一，后任教于北平京华美术专科学校。

　　管平湖先生对古琴艺术研究极深，得九嶷派杨宗稷、武夷派悟澄老人及川派秦鹤鸣等名琴家之真传，他能博取三派之长，并从民间音乐中吸取营养，融会贯通，不断创新，自成一家，形成近代中国琴坛上有重要地位的"管派"。其传人有袁荃猷（1920—2003）、郑珉中（1923—2019）、王迪（1923—2005）、许健（1923—2017）等，田双坤（1933—）是其唯一斫琴弟子。

管平湖先生将毕生精力倾注于古琴事业，不仅演奏技艺精湛，闻名于世，而且琴学造诣极深，特别在发掘古谱方面，做出了巨大的贡献，如《广陵散》《碣石调·幽兰》《离骚》《大胡笳》《胡笳十八拍》《秋鸿》《欸乃》等许多著名古琴大曲，均由他率先发掘打谱，通过他的艰苦努力，这些绝响已久的古琴曲得以重新恢复了艺术生命，他对古琴工作有起潜振绝的雄伟功绩。1977年8月20日，美国发射的旅行者号卫星，将管先生演奏的古琴曲《流水》录入喷金的世界名曲唱片中，使中国古琴第一次响彻太空。他所著《古指法考》一书，对指法研究提出了他自己的真知灼见。

管平湖先生斫仿各式古琴十余张，修琴无数。他曾为故宫博物院修理清宫旧藏唐"大圣遗音"琴，清除了数十年间因屋漏流淌琴面的泥水锈污，使之恢复了金徽，呈现出琴面的本来面目，而无丝毫的伤损。据郑珉中先生讲："管先生平日为门人修琴很多，如李伯仁旧藏的'飞泉'琴，程宽初得时发音异常细微，经先生修后再现了温润而雄厚之音。蔡氏得连珠式琴，而发音欠圆，先生将其改为'独幽'的形式后，解决了按音不圆的问题。"

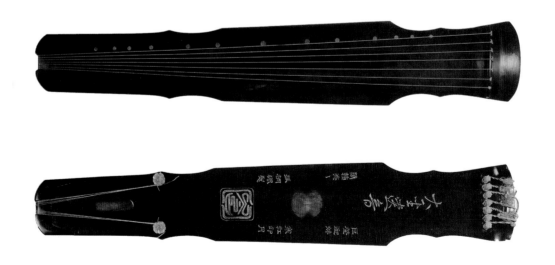

管平湖修唐神农式"大圣遗音"琴　北京故宫博物院藏

Guan Pinghu （1897—1967）

Given name Ping, courtesy name Ji'an and Zhongkang, and art name Pinghu, came from Suzhou, Jiangsu Province(now Suzhou, Jiangsu). He claimed himself to be a layman of Guqin. He is the son of Guan Nianci, a famous court painter in the Qing Dynasty. From an early age, he had developed a keen interest in art. Therefore, he studied qin and painting from his family, and paintings of flowers and figures from Jin Shaocheng, a well-known painter. He was particularly good at traditional Chinese realistic painting, with a beautiful and novel vigor of style, setting the well-established rules aside. He was a core member of the Hu Society Painting Association and later taught in Beiping Jinghua Academy of Fine Arts.

Mr. Guan Pinghu has probed deep into the art of Guqin and studied from famous qin players such as Yang Zongji of Jiuyi School, Old Man Wu-cheng of Wuyi School and Qin Heming of Sichuan School. Drawing on the strengths of the three schools, while draining folk music for nourishment, he had developed a style of his own, forming the "Guan School" with an important position in the qin world in modern China. His students include Yuan Quanyou (1920-2003), Zheng Minzhong (1923-2019), Wang Di (1923-2005), Xu Jian (1923-2017) and others. Tian Shuangkun (1933-) is his only student who engaged in qin making.

Mr. Guan Pinghu devoted his whole life to the art of Guqin, which made for his superb skills, widespread reputation, and profound attainments in qin studies. Especially, he has made great contributions to the exploration of ancient Guqin notations, such as Guangling Melody, Secluded Orchid in Jieshi Mode, The Sorrow of Parting, Great Normad Reed-pipe, Eighteen Songs of a Nomad Flute, Wild Swans of Autumn, and Boatmen's Chant, etc. Through his concerted efforts, these long-standing Guqin classics regained their artistic vigor. On August 20, 1977, the Voyager satellite launched by the United States recorded Mr. Guan's Flowing Water into the gold-sprayed world-famous music record, making the timber of Chinese Guqin heard by the space for the first time. His book Study on Ancient Fingering provides his own penetrating insights into the study of fingering.

Mr. Guan Pinghu made ten-odd qins by referring to ancient qins of various styles and repaired countless of them, including the "Musical Legacy of the Sage" collected in the Palace Museum. By removing the mud and rust on the qin surface due to roof leaking for decades, he restored the true colors of the jade marker and the qin surface without doing any damage. According to Mr. Zheng Minzhong, "Mr. Guan repairs a lot of qins for his disciples. For example, when Cheng Kuan acquired the "Feiquan" qin once collected by Li Boren, he found the sound it produced was extremely subtle, so he resorted to Mr. Guan. After being repaired by Mr. Guan, the qin regained its gentle yet strong timbre. Mr. Cai got a Lianzhu (linked bead) style qin, but only to find its sound rough. After being changed into the "solitary seclusion" style by Mr. Guan, the problem was solved.

田双坤

　　田双坤，时用名"田双琨"，字双魁，1933年出生于河北省深县（今河北省深州市）。1951年，他在北京琉璃厂马良正乐器店学徒，拜马文元学习制作京胡。1956年该店并入北京民族乐器厂后，他负责制作板胡、二胡等。1972年，他开始担任艺术界和乐器厂联合成立的"琴、筝、瑟"乐器改革小组长，积累了丰富的乐器制作和改良经验。传人有三百余人。

　　1958年田双坤先生师从中国著名古琴演奏大师、制作大师管平湖先生学习斫琴，是管平湖先生唯一斫琴弟子。他跟随管平湖先生第一次制作的三张琴在北京乐器厂举办的鉴定会上获得著名琴家溥雪斋、查阜西、吴景略、管平湖等人的一致好评。

　　时至今日，田双坤先生已经88岁高龄，斫琴授徒，深耕不辍，一生斫琴修琴无数。其斫制的古琴得到同行高度认可，深受诸多知名琴家喜爱并被他们收藏，他是北方最有声望的斫琴大家，被尊称为"古琴田"。他担任中国古琴学会顾问、中国民族器乐学会理事等职，《中国乐器大全》《中国古琴民间典藏》《古琴荟珍》《中国乐器博物馆》《手艺北京》等著作对田双坤先生斫制古琴、修复古琴的贡献均有记载。2018年，在田双坤先生从事斫琴事业60周年之际，中国民族器乐学会、北京乐器学会授予其"中国民族乐器终身成就奖"。

Tian Shuangkun

Current name "Tian Shuangkun (琨)", born in 1933 in Shenxian County, Hebei Province. In 1951, he was an apprentice of Ma Liangzheng Musical Instrument Store in Liulichang, Beijing, learning how to make Jinghu (a two-stringed bowed instrument with a high register) from Ma Wenyuan. In 1956, he was hired by the Beijing National Musical Instrument Factory to make Banhu (a bowed two-stringed instrument with a thin wooden soundboard) and Erhu. In 1972, he served as the leader of the "Qin, Zheng and Se" musical instrument reform team jointly established by the art world and the factory, accumulating a wealth of experience in musical instrument making and improvement. He has more than 300 disciples.

In 1958, Mr. Tian Shuangkun began to learn to make qin from Mr. Guan Pinghu, a renowned Chinese master of Guqin studies and master maker of qin. He was the only one who learned qin making from Mr. Guan Pinghu. The three qins made by him for the first time, under the guidance of Mr. Guan, won unanimous praise from famous qin players Pu Xuezhai, Zha Fuxi, Wu Jinglue, Guan Pinghu and others at an appraisal session held by Beijing Musical Instrument Factory.

Almost in his 90s though, Mr. Tian Shuangkun is sticking to student teaching and qin making. Throughout his life, he has made and repaired countless qins. Qins made by him are highly recognized by his peers and are collected by many well-known Guqin artists. He himself also becomes the most prestigious master maker in north China and is hailed as "Guqin Tian". He currently serves as a consultant of China's Academy of Guqin and a director of the Chinese Society of National Instrumental Music. Mr. Tian Shuangkun's contribution to Guqin making and restoration is included in such works as the Treasury of Chinese Musical Instruments, the Folk Collection of Chinese Guqin, the Guqin Huizhen, the Chinese Musical Instruments Museum and the Handicraft Beijing. "Chinese Guqin", "Chinese Folk Music", "Cultural Geography", "Cultural Monthly", "Vogue", "Top China", "Divineland", "China Today" and CCTV all have done feature stories on him. In 2018, at the 60th year of Mr. Tian Shuangkun's engagement in qin making, the Chinese Society of National Instrumental Music and the Beijing Musical Instrument Society awarded him the "Lifetime Achievement Award for Chinese National Musical Instruments".

昭　闻

　　昭闻，名程刚，字昭闻，研究员，中国民主建国会会员。陕西省非物质文化遗产古法斫琴代表性传承人，北派古琴制作代表人，中国古琴博物馆馆长，中国音乐家协会会员，中英国际音乐节中国区主席，陕西省长安古琴艺术研究院院长，陕西省技能大师，西安市高层次领军人才，西安市政府职业能力专家，西安市民族管弦乐学会会长，西安琴会会长。他常年致力于以纯手工古法斫琴以及长安古琴艺术史研究，发表学术论文有《长安古琴艺术考》《古琴专业音乐教育的历史与现状》《古琴文化中的和合思想探微》《古琴琴歌的艺术特征分析》《论中国古琴文化精神内涵的传承及当代意义下的审美重构》等十余篇。古琴历史专著《长安古琴艺术》由陕西人民出版社出版发行，荣获西安市非物质文化遗产专项资金

资助、西咸新区沣东新城优秀文艺作品专项资金资助。

昭闻出身医学世家，从北京中医药大学毕业后，源于对中国传统文化和古琴艺术的热爱，曾拜入九嶷派第二代传人韩廷瑶老先生门下学习抚琴，2014 年追随田双坤先生学习管派古琴制作技艺，2018 年执拜师礼，深得管派真传，并得到各地古琴大师与斫琴大家的悉心指导。他创立了中国古琴博物馆、陕西省长安古琴艺术研究院（研究院传习中心在册学员一千余人），主办国家级、省级音乐会、研讨会 30 余场次，是西北地区古琴事业发展的领军人。其古琴制作技艺现有传人姚帧、程博源等。

昭闻古法斫琴以工艺复杂、时间漫长、对原材料苛刻而著称。每张琴均采用百年以上古旧庙梁为原材料，使用秦岭产苎麻通体包裹，梅花鹿角霜调和秦岭野生生漆作为灰胎，遵照传统工序，耗时两年以上斫制而成，其音色高古、通透、圆润，余韵悠长，低音醇厚、高音清越，可与唐宋老琴媲美，是北方古琴的典型代表。

Zhao Wen

Autonym Cheng Gang,courtesy name ZhaoWen, member of the China Democratic National Construction Association (CDNCA). The representative inheritor of Shaanxi Intangible Cultural Heritage in Guqin Making with Ancient Techniques; representative of Beipai Guqin Production; curator of Chinese Guqin Museum; member of the Chinese Musicians Association; Chairman of China Region of China-UK International Music Festival; dean of Shaanxi Chang'an Guqin Art Institute;Skill Master of Shaanxi Province; high-level leading talent of Xi'an; professional ability expert of Xi'an Municipal Government.President of Xi'an National Orchestra Association; and the President of Xi'an Qin Association.He has devoted himself to the study of pure handmade ancient qins and the history of Chang'an ancient qins for years. He has published more than ten academic papers such as Research on Chang'an Guqin Art, History and Current Situation of Guqin Professional Music Education, Exploration of Harmony Thought in Guqin Culture, Analysis of Artistic Characteristics of Guqin Songs, and On Inheritance of Spiritual Connotation of Chinese Guqin Culture and Aesthetic Reconstruction in Contemporary Context. His historical monograph Guqin Art in Chang'an was published and distributed by Shaanxi People's Publishing House, funded by Xi'an Intangible Cultural Heritage Special Fund and the Special Project for Outstanding Literary Works in Fengdong New Town, Xixian New District.

Coming from a medical family, ZhaoWen, upon graduating from Beijing University of Chinese Medicine, began to study qin playing from Mr. Han Tingyao, the second-generation successor of Jiuyi School, out of his passion for Chinese traditional culture and Guqin art. In 2014, he followed Mr. Tian Shuangkun to learn to make Guan-school Guqin. The teacher worship ceremony was held in 2018. He has learned the essence of Guan-school Guqin making skills and was counseled by renowned Guqin masters and master makers nationwide. He set up the Chinese Guqin Museum and the Shaanxi Chang'an Guqin Art Institute, which, with more than 1,000 registered students, has hosted 30-odd national and provincial concerts and seminars. As the leader in the development of the Guqin art in Northwest China, his disciples in Guqin making include Yao Zhen and Cheng Boyuan.

ZhaoWen's ancient qin-making skill is famous for its complicated technique, long duration, and harsh demand on raw materials. Each qin is made of more than 100-year-old temple beams, wrapped in full-body by ramie grown in Qinling Mountains, pasted with lacquer cement made by blending cornu cervi degelatinatum of spotted deer and wild raw lacquer from Qinling Mountains and then assembled in accordance with traditional procedures. The whole process takes more than two years. Its timbre is loud, transparent, round, and long-lasting, coupled with mellow bass and clear treble. As a typical representative of ancient qins in north China, it can be even on a par with the old ones made in Tang and Song Dynasties.

第一章 工具

Chapter I
Tools

直尺

主要用于测量琴体长、宽、高
等尺寸。

Ruler: mainly used to measure
the length, width and height of
the qin body.

角尺

主要用于画线、画角等。

Steel square: mainly used
for drawing lines, angles,
etc.

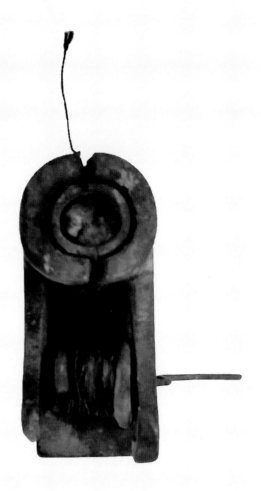

墨斗

用于放线。

Chalk line: used for marking long, straight lines.

靠尺

主要用于琴胚和灰胎施工中的找平参考。

Guiding rule: mainly used for leveling in qin plank and grey gatch construction.

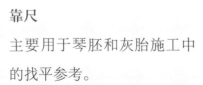

锯

主要用于裁切木板、
配件等。

Saw: mainly used for
cutting wood boards,
accessories, etc.

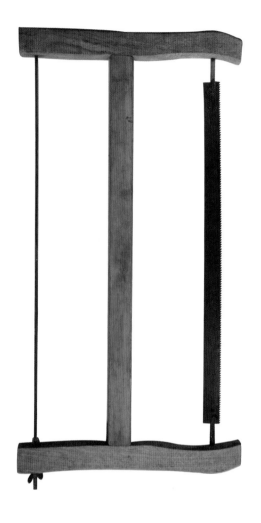

开凿锯

主要用于安装岳山等需
要开凿的地方。

Digging saw: mainly used
to install bridge and other
parts that need digging.

斧

主要用于砍出琴体的大概形状
和琴面的大致弧度。

Axe: mainly used to cut out the
approximate shape of qin body
and construct the approximate
radian of the surface.

刨

分为长刨、短刨、小刨等，主要用于琴体表面修整。

Plane: classified into long plane, short plane, small plane, etc. It is mainly used for surface trimming of the qin body.

短刨

用于刨出琴面弧度和低头。

Short plane: used to plane out the radian of qin surface and the lower end.

长刨

用于琴体边墙的找直。

Long plane: used to straighten the side wall of the qin body.

小刨

用于琴体槽腹内部修整。

Small plane: used for trimming the interior of the groove of the qin body.

锉

分为木锉和钢锉，木锉主要
用于琴体的精细修形，钢锉
主要用于红木配件的打磨。

File: classified into wood file and
steel file. Wood file is mainly
used for detail modification
of the qin body, and steel file
is mainly used for polishing
mahogany accessories.

板锉

主要用于冠角、岳山等有弧度的
红木表面的修整。

Plate file: mainly used for trimming
the mahogany surfaces with radian
such as crown angle and bridge.

蜈蚣锉

主要用于红木配件外表的平整
光滑处理。

Centipede file: mainly used to
smooth the surface of mahogany
accessories.

铲

分为平铲和圆铲。

Shovel: divided into flat shovel
and round shovel.

平铲

主要用于琴体及配件的安装。

Flat shovel: mainly used for
the installation of qin body and
accessories.

圆铲

主要用于琴体槽腹的挖制，分
为大、中、小三种型号，大号
用于粗挖，中、小号用于后期
对槽腹进行调整。

Round shovel: mainly used for
digging the belly groove of the
qin body. It is divided into three
types: large, medium and small.
The large is used for rough
digging, and the medium and
small are used for adjusting the
belly groove later.

钻

主要用于打孔、制作绒扣眼等。

Drill: mainly used for drilling
holes, velvet buttonholes, etc.

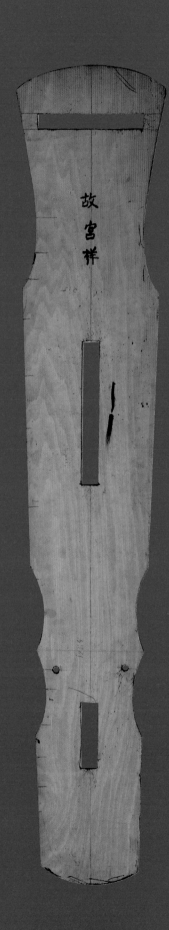

伏羲式模板

伏羲，华夏民族的人文始祖，三皇之一，相传中国文化的许多发明与创造都与他有关。据说古琴即由伏羲发明，《琴书》《长笛赋》中都有"伏羲削桐为琴""昔疱羲作琴"的记载。伏羲式琴造型浑圆、古朴，项、腰各有半月形弯入，音色宽宏。此伏羲式为北京故宫博物院藏"九霄环佩"古琴样式。

Fuxi style template: Fuxi, the human ancestor of the Chinese nation, ranks first among the Three Sovereigns. Many inventions and creations of Chinese culture are related to him. It is said that Guqin was invented by Fuxi, and there are records in Qin Shu and Flute Rhapsodies that "Fuxi carved tung trees into qin" and "Paoyi(Fuxi) qin long time ago". Fuxi style qin is round in shape, simple and unsophisticated, with a half crescent-shaped bend in the neck and waist, and a deep and rich timbre. This Fuxi style is the "Jiuxiao Huanpei" Guqin style collected by Beijing Palace Museum.

仲尼式模板

"仲尼式"又称"夫子式"，孔子曾学琴于师襄子，后按照自己的理想设计琴式，于是，后人称孔子创造的琴式为"仲尼式"。仲尼式在项、腰处各呈方折凹入，造型简洁朴素，声音清雅纯正。此仲尼式模板为1973年5月29日由双坤先生根据1958年夏天管平湖先生提供的清代仲尼式古琴而制作。

Zhongni style template: "Zhongni style" is also called "Confucius style". Confucius studied qin from Shi Xiangzi, and later made a qin based on his own ideal and designed certain rules. Later generations called the qin style created by Confucius "Zhongni style". The Zhongni style is folded and concave at the qin's neck and waist, with the simple shape and pure sound. This Zhongni style template was made by Mr. Tian Shuangkun on May 29, 1973 according to the Qing Dynasty Zhongni tyle Guqin provided by Mr. Guan Pinghu in the summer of 1958.

落霞式模板

落霞式据说是文人在傍晚观察千变万化的晚霞时迸发出了灵感，并据此斫制成琴。相传"落霞"是古代名琴之一，汉郭宪《洞冥记》卷三中有曰："握凤管之箫，抚落霞之琴。"此落霞式模板为管平湖先生所制"大扁儿"古琴样式，琴体相对宽大。

Sunset style template: Sunset style is said to be inspired by literati's observation of the ever-changing sunset glow at dusk and the inspiration is imbued into a qin. "Sunset" is one of the famous qins in ancient times. A line in Volume 3 of Guo Xian's Dong Ming Ji in Han Dynasty goes like this: "Hold the flute with the phoenix pipe and play the qin at sunset". This sunset template is the style of "Da Bian Er" Guqin made by Mr. Guan Pinghu, and its body is relatively wide.

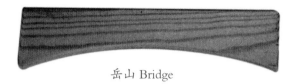

岳山 Bridge

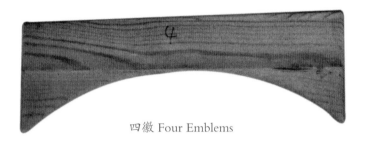

四徽 Four Emblems

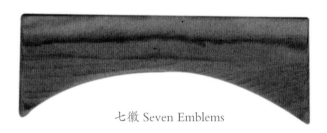

七徽 Seven Emblems

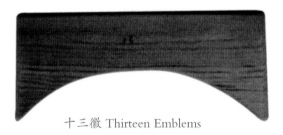

十三徽 Thirteen Emblems

琴面弧度范

是用于规范岳山、四徽、七徽、十三徽四个部位琴面弧度的卡尺，参照 1958 年夏天管平湖先生挑选的音色、手感、外观俱佳的清代仲尼式古琴制作而成。

Radian of qin surface: the radian caliper mainly composed of bridge, four emblems, seven emblems and thirteen emblems made with reference to the Qing Dynasty Zhongni style qin with good timbre, hand feeling and appearance selected by Mr. Guan Pinghu in the summer of 1958.

碾槽

用于碾碎鹿角霜、八宝灰等。

Grinding groove: used to grind
antler cream, eight treasures
ash, etc.

臼

用于捣碎朱砂、藤黄等与生
漆配合的矿石类颜料。

Mortar: used to mash cinnabar,
gamboge and other mineral
pigments mixed with raw
lacquer.

剪刀、美工刀

主要用于裁剪葛布和修整已裹好的葛布等。

Scissor and box cutter: mainly used to cut ko-hemp cloth, trim wrapped ko-hemp cloth, etc.

刮刀、刮铲

主要用于粗、中、细灰胎的施工。

Scraper and spatula: mainly used for the construction of coarse, medium and fine grey gatch.

橡皮刮刀

主要用于琴体上面漆前的砂眼、毛孔闭合的施工。

Rubber spatula: mainly used for the construction of sand holes and pore closure before painting on the qin body.

60 目
60 mesh

120 目
120 mesh

800 目
800 mesh

2000 目
2,000 mesh

砂纸

由粗到细，用于干燥后的灰胎或
面漆打磨。

Sandpaper: with coarse to fine
specifications, used for grinding
dry gray gatch or topcoat.

砂纸枕

由四边平直木方制成，裹上砂纸
用于打磨琴体。

Sandpaper pillow: made of straight
square wood on four sides, wrapped
with sandpaper for polishing the
qin body.

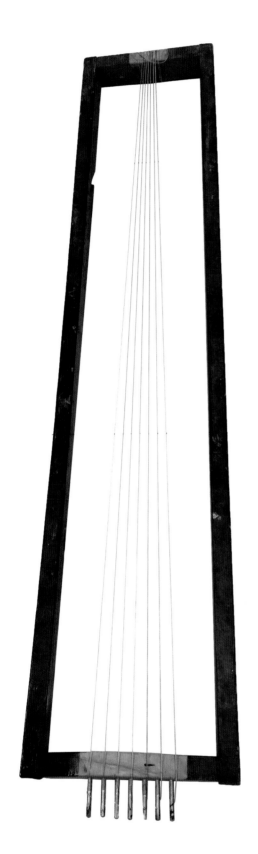

试音弓

用于从琴体木胚开始至粗灰、中灰、细灰施工期间的试音、排煞音等。

Audition bow: used for audition and braking during the construction from the wooden embryo of the qin workplank to coarse ash, medium ash and fine ash.

刻刀

主要用于雕刻焦尾、龙龈等红木配件。

Carving knife: mainly used to carve mahogany accessories such as Jiaowei and Longyin.

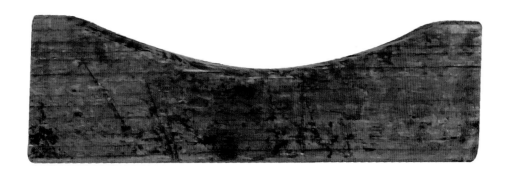

弧形枕

斫琴中用于保持琴体的稳定。

Arc pillow: used to maintain the stability of the qin body during the construction process.

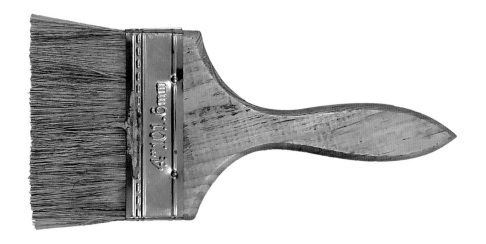

排刷

主要用于靠木漆及底漆的施工。

Brush row: mainly used for the construction of wood paint and primer.

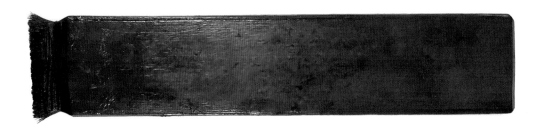

发刷

主要用于面漆的施工。

Hair brush: mainly used for topcoat construction.

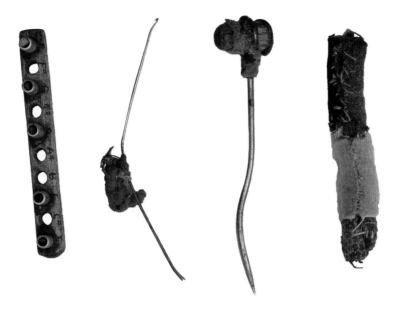

上弦套组

主要由定音哨、穿绒扣器、小尖钻、拉弦器组成。

Upper string sleeve group: mainly composed of sound fixing whistle, velvet threading buckle, small sharp drill and string pulling device.

定音哨

由能吹出各类常用标准音高的铜哨组成，上弦时作为琴弦音高的参考。

Sound fixing whistle: composed of various copper whistles with standard sound heights and is used as a reference for string pitch when winding.

第二章 选材

Chapter II
Selection of Materials

　　古琴制作一般多选用桐木、杉木等纹理顺直、无节疤的老旧材料，本次用料为 100 年老杉木。

　　Guqin is usually made of aged materials with uniform texture and no knots, such as paulownia wood and Chinese fir wood. This time, the material used is 100-year-old Chinese fir wood.

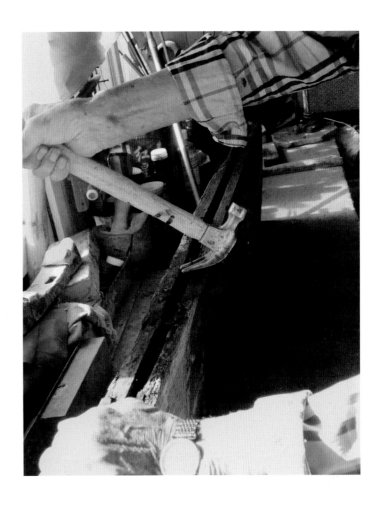

检查老旧材料中是否有破损，铁钉等。

Check the old materials for damages, nails, etc.

用斧初步去除腐
烂、开裂等边部不能
使用和妨碍刨子找平
的部位。

The axe is used to
preliminarily remove
the rotten, cracked parts
that cannot be used or
hinder the plane from
leveling.

根据材料的综合情况选择可制作的古琴样式，并提前确定材料适合的正反面和高低音位置等。一般来说用面板材料纹理凸出面作为琴面，凹陷面作为槽腹。纹理疏松部位留作大弦位置（外），纹理紧密部位留作小弦位置（内）。靠树梢位置留作琴尾，靠树根位置留作琴头。

Based on the comprehensive status of the materials, select a feasible Guqin style, the suitable front and back sides, high and low pitch positions in advance. Generally speaking, for panel material, the "convex" side is used to make the qin surface and the "concave" side is to make the groove belly.The part with loose texture is reserved for the large string position (outside), and the part with tight texture is reserved for the small string position (inside). The treetop part of the wood is reserved for the tail of the qin and root part is reserved for the head of the qin.

本次老杉面板材料应作为琴弦一面的"凸"纹之上有一条深达1厘米的通体裂缝。为了避免制成琴面后出现开裂为两半的情况，故以通体裂缝一面，也就是前文中所说的"凸"纹理一面作为槽腹的部位。

This time, however, there is a crack up to 1cm deep running through the string part of the Chinese fir panel with "convex" texture. To prevent the qin surface from cracking into two halves, the crack side of the material, that is, the side with "convex" texture mentioned earlier, is taken as the groove belly.

　　用靠尺作为参照依据，分别对面板长、宽进行找平处理。

　　The length and width of the panel are leveled respectively using the guiding rule as the reference basis.

古琴制作，原材料的选择是首要的，正所谓"巧妇难为无米之炊"，再好的斫琴师如无好的原材料，也施展不开自己的本领，表现不出自己的水平。

古琴选材，古人总结为四善——"轻""松""脆""滑"。轻是指木材重量要轻；松是指选用的木材要松透；脆是指材料有脆性，敲击声音清脆、响亮，易折断；滑是指用料经过打磨后顺直、光滑。

斫琴的原材料讲究用老木头，越老越好，但是不能过度腐朽、掉粉末，即"老而不朽"。老木头制琴无火气，比较稳定，不易变形、开裂。

老木头多用老房梁、门板等。也有人用棺椁，棺椁以悬棺为良。埋入地下的棺椁长年吸收地气，阴气太重，声音沉闷，而且容易开裂，最好进行返阳处理，需放置六七年时间，因此个人不建议使用棺椁做琴。

古琴的底板材料一般比面板硬一些，也有的是用与面板相同的材料，称之为纯阳琴，但要根据音色情况用生漆灰胎对底板进行适当的处理。

面板多选用杉木、桐木、松木等，总体来说要选择纹理顺直的木材，树木中段最好，因为近树根处的木材音浊，不够松透、清脆，近树梢处音易空泛、易飘。但是一个好的斫琴师也可以利用这些特点，斫出更好的古琴，高手斫琴不拘泥于材料。木质硬重的材料在制作时刮薄一点，木质松软的材料相应厚一点。旧时也有人用樟木等硬杂木制琴，音色也很不错。现在有年代的老木材越来越少，部分依靠流水线大批量生产古琴的厂家多用经过烘烤的新木材。烘烤过的木材做出的琴音质噪，因为烘烤过程中木材的纤维已经断裂、破坏，没有自然风干的木材完整，所以还是自然风干的木材音色更好一些。

底板多用梓木、枫木、色木、松木、金丝楠木、椿木，它们与桐木、杉木相配发声很好。梓木是做琴传统用料，北方称为楸木，性能稳定，不易开裂。枫木和色木也是制作提琴底板的传统材料，用来制作古琴底板音色很好，但色木容易变形，需要处理好。松木品种比较多，硬度不一样，白松比较软，白花松最硬。椿木有香椿、臭椿、白椿三种，用香椿、白椿制作底板，声音都很不错。

总的来说，无论面板、底板，可使用的材料很多，面底材料的软硬度比十分重要。

Raw material is what count most in making a Guqin. As the saying goes, "It is difficult for a skillful woman to cook without rice". No matter how skillful a master maker is, he cannot display his abilities or show his level without good raw materials.

The ancients summed up the good qualities of Guqin as "light, loose, crisp and smooth". Light means that the wood should be in light weight; Loose means that the texture of the selected wood should be loose; Crisp means that the material is solid, the sound of knocking on it is crisp, and it is easy to break; Smooth means that the material should be straight and smooth after polishing.

The raw materials for qin making should be old wood, the older the better. But they should not decay excessively or crumble. In other words, they are old but still firm. The qin made of old wood is relatively stable, and not prone to deformation or crack.

Old wood are usually selected from used beams, used doors, etc. Coffins are also used, among which hanging coffins are the prime. Coffins buried underground absorb the underground air for many years. The yin air is too heavy, making the sound dull and the wood easy to crack. It is better to carry out anti-yin treatment and leave it untouched for six or seven years. Personally I advise against using the coffin wood to make qin.

The bottom panel material is generally more solid than that of the top panel. When the same material is used for both panels, the qin is called Pure Yang Qin. However, the bottom panel should be properly treated with raw lacquer grey patty consistent to the timbre.

The top panel is mostly made of Chinese fir, paulownia wood, pine wood, etc. Generally speaking, the texture should be uniform. Therefore, the middle section of the tree is the best, because the wood near the root makes the sound of qin dull-not loose or crisp enough, and the wood near the treetops makes qin's sound empty and floating. However, a good master maker can also make use of these characteristics to produce better Guqin. In other words, material itself may not be so important for a good master maker. Hard and heavy wood materials should be processed thinner, while soft wood materials thicker accordingly. Back in the day some people also made qin from hardwood such as camphor wood, and the timbre was also very good. There are fewer and fewer old natural timber to be used for making qin now, and some manufacturers with assembly lines mostly use dried timber to make qin. The sound quality of the qin made of the dried wood is coarse, because the fibers in the dried material have been broken and destroyed, and the texture is not as good as naturally dried wood. Therefore, the timbre of qin made of naturally dried wood is better.

The bottom plate is mostly made of ovate catalpa wood, maple wood, strained wood, pine wood, phoebe zhennan wood and toon wood. Combined with paulownia wood and fir wood, they make the sound of qin excellent. Ovate catalpa wood is a traditional material for making qin. It is called qiu wood in the north. It has stable quality and is not prone to crack. Maple and strained wood are traditional materials for making qin's bottom panel. They make the timbre of qin very good. Strained wood is easy to deform and must be handled properly. There are many varieties of pine wood with different hardness. White pine wood is soft and white flower pine wood is the hardest. There are three kinds of chun wood: Chinese mahogany, varnish and white mahogany. Chinese mahogany and white mahogany are perfect for making the bottom panel.

In general, whether it's the top panel or the bottom panel, there are many materials available,and the softness-hardness ratio of the material for the surface and bottom panels should be taken into account.

第三章　造型

Chapter III Modeling

　　根据材料的特点和自己对古琴形制的喜好选择古琴的样式，本次面板材料较好、较宽，在仲尼式、伏羲式、落霞式三种古琴样式中，最终选择故宫藏唐代伏羲式"九霄环佩"琴式。

According to the characteristics of the materials and personal preference, the style of Guqin is selected. This time, the panel material is better and wider. Among the three types of qin-- Zhongni style, Fuxi style and Sunset style, the Fuxi-style "Jiuxiao Huanpei" of the Tang Dynasty, which is enshrined in the Forbidden City, is finally selected.

图为本次挑选的面板的特写，可以看
到有很多蛀虫留下的小孔。

In the close-up of this selected panel, there
are many small holes left by moths.

按照模板用墨签画出琴样。

Draw the qin contour with ink based on the template.

按照画好的墨线锯除多余
的部分。

Saw off the excess part along
the ink line.

锯好面板形状以后画出面板预留的厚度，一般面板先预留 1 厘米厚度，预留出以后修整的余地。

After sawing out the contour of the panel, mark the reserved thickness of the panel. Generally, the panel is reserved with a thickness of 1 cm first, leaving enough room for later trimming.

用斧头逐步砍出大致的弧度。

Use the axe to cut out the approximate radian step by step.

大致砍出低头。

Cut out a preliminary lower side head.

用短刨进一步对砍好的琴面进行修整。

Use the short plane to further trim the cut qin surface.

标出四、七、十三徽的位置，以便使用琴面弧度范进行测量。

Mark the positions of the 4th, 7th and 13th emblems so as to use the radian of the qin surface for measurement.

标出琴尾的厚度，一般故宫藏唐琴的厚度在 3.8 至 4 厘米之间。

Mark the thickness at the tail of qin. Generally, the thickness of Tang Qin in the Forbidden City is between 3.8 and 4 cm.

进一步用刨子刨，并用手抚摸已刨好的琴面，感受弧度是否平整及其舒适度。

Further plane it and touch the planed surface with your hand to feel the smoothness and texture of the radian of the surface.

　　用琴面弧度范按照已标好的四、七、十三徽的位置测量琴面弧度。

　　Measure the radian of the surface based on the positions of the marked 4th, 7th and 13th emblems on the qin surface.

用模板画出岳山的位置，再用岳山位置的弧度范确定岳山位置的弧度。

Mark the position of bridge on the template, and then use the radian ruler to measure the radian at the position of the bridge.

用弧度范对琴面反复测量，再用刨子反复修整，使其达到平顺、光滑，各部位弧度合适为止。

Use radian ruler to measure the qin surface repeatedly, and then use plane to trim it repeatedly to achieve smoothness until the radian of each part is appropriate.

　　用长板尺和短尺90度交叉测量琴头的低头尺寸，从高点到岳山位置垂直距离为1.2厘米。根据弧度范和低头的尺寸要求，再次检查并用刨子精修，达到满意为止。

　　When measuring the size of the lower side of the qin head, an intersection angle of 90° between the long board ruler and the short ruler is applied. The vertical distance from the high point to the bridge is 1.2 cm. According to the size requirements of radian ruler and lower side head, check again and finish with plane until it is satisfactory.

　　根据面板预留的画线对面板内外两侧
使用刨子进行找直精修。

　　According to the lines drawn on the
panel, the inner and outer sides of the panel are
refined with planes.

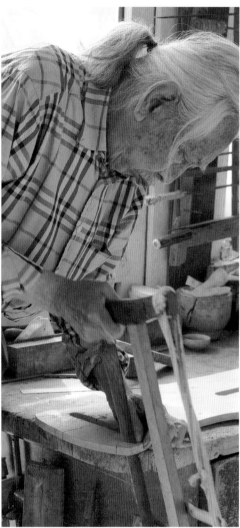

用模板画出底板的样子，再用曲线锯裁切，用钻、锯等对龙池、凤沼、轸池进行开孔，锯好安装护轸的预留位置。

Draw the contour of the bottom plate with a template, then cut it out with a curved saw, and make holes at Longchi, Fengzhao and Zhenchi with drills and saws.Reserve the position for the installation of the Huzhen using saw.

用各种合适型号的锉，
对龙池、凤沼、轸池进行修
整，直到平整光滑。

Use various types of files
to trim Longchi, Fengzhao and
Zhenchi until they are smooth.

用弧度锉锉出龙池、凤沼的圆弧。

Use radian file to file out the circular arcs of Longchi and Fengzhao.

对底板外侧两边二至三指宽的位置用刨子修薄。

Reduce the thickness by 2 to 3 fingers on both sides of the outer side of the bottom panel with a plane.

古琴的属性首先是一种乐器，能发出完美的声音是第一要求，同时也要有一个漂亮的外观，而外观与音色往往是成正比的。

这次的杉木琴材十分难得，不但古旧，而且四善俱全，虫蛀程度也恰到好处，这种材料斫出的琴音色古朴、松透、洪亮。根据此材料的尺寸尽量选择比较宽大的形制，三种模板中选择故宫样（北京故宫博物院藏唐代伏羲式"九霄环佩"古琴），就是为了充分发挥材料的优势。

唐代伏羲式"九霄环佩"琴，通长124厘米、额宽21.8厘米、肩宽21.2厘米、尾宽15.4厘米、厚5.8厘米，此琴形制特点是厚重、边角浑圆。造型如果处理得当，琴型古朴、浑厚而大方，如果处理不得当，会给人一种粗鄙、笨重的感觉。制作时需要将项与腰内收部位上下边做圆，让棱角线向内移，缩小与上下侧面的厚度差，琴头额部渐薄，达到匀称的良好视觉效果。这些在《故宫古琴图典》中都有记载。

每一个时代有每一个时代的审美标准，"唐圆宋扁"是唐代和宋代的审美标准，在那个时代这个标准就是最实用的、最美的，但是以现在的审美标准来看，总体来说，"圆"与"扁"都有一些过了，如果不是文物复制需求，还是以现代审美和使用习惯为标准最好。此琴制作时建议琴边的厚度适当薄一点，琴面弧度使用管平湖先生1958年精选的音色、手感、外观俱佳的琴面弧度范本，在保存传统核心技艺的前提下，适应现代审美和使用习惯。

Guqin is first of all a musical instrument, so the ability to produce sound matters most. At the same time, it is better to have a beautiful appearance, which is often proportional to the timbre.

This time, the fir wood material is rare. It is not only aged, but also has all the four good qualities and is properly moth-eaten. The timbre of the qin made of this material is loud, clear and resonant. According to the size of this material, it's better to choose a larger configuration. Among the three templates, the Imperial Palace template is chosen (the Fuxi-style "Jiuxiao Huanpei" qin of the Tang Dynasty enshrined in Beijing Palace Museum) in order to give full play to the advantages of the material.

The Tang Dynasty Fuxi-style "Jiuxiao Huanpei" qin has a length of 124 cm, a forehead width of 21.8 cm, a shoulder width of 21.2 cm, a tail width of 15.4 cm and a thickness of 5.8 cm. The shape of the qin is characterized by massiness and roundness. If the modeling is properly handled, the qin will appear simple, unsophisticated, vigorous and generous. If not, it will give people a vulgar and cumbersome feeling. Its appearance is difficult to be perfect. It is necessary to round the upper and lower edges of the items and waist adduction parts, so as to move the cornering angle line inward, reduce the thickness difference with the upper and lower sides, and reduce the thickness at the head of the pin, thus achieving a well-proportioned visual effect. These are all recorded in Classics of the Forbidden City: Guqin in the Collection of The Palace Museum.

Each era has its own aesthetic standard. As a classical saying goes, Tang-dynasty qin were round, and Song-dynasty ones flat. In those eras, items meeting these standards were seen as practical and beautiful. However, to this day, the "round" and "flat" have gone outdated on the whole. If it is not the demand for cultural relics reproduction, it is better to adopt modern aesthetic and habit of usage as the standard. When making this qin, it is suggested that the side wall should be thinner. The radian of the qin surface should refer to the radian model selected by Mr. Guan Pinghu in 1958, which is of good timbre, hand feeling and appearance. On the premise of preserving the traditional classic skills, it should be adapted to modern aesthetics and habit of using.

第四章 槽腹

Chapter IV
Ventral Groove

　　用墨签画出琴腹边墙，边墙厚度内侧

（七弦）为 0.6 厘米，外侧（一弦）为 0.8

至 1 厘米。

　　Draw the ventral wall of the qin with
ink sticks. The thickness of the wall is 0.6
cm on the inner side (seven chords) and 0.8-
1cm on the outer side (one chord).

琴头位置与轸池预留 1 厘米宽度。

Reserve 1 cm width between the position of the qin head and the zhenchi.

琴尾边墙预留厚度为 3 至 6 厘米。

The reserved thickness of the wall of the qin tail is 3-6 cm.

用模板画出龙池、凤沼、雁足位置，将画好的龙池、凤沼的位置长度延长 2 厘米，宽度增加 1 厘米，雁足沿中心点画半径 2 厘米半圆作为预留，用大铲刀对琴腹挖制，注意在挖制中不要伤及预留部分，龙池、凤沼部位的留实采用斜坡向下（上小下大）的方式，雁足预留部分采用垂直向下的方式挖制。

Draw the positions of longchi, fengzhao and yanzu based on the template. The length of the drawn longchi and fengzhao is extended by 2cm and the width is increased by 1cm. The yanzu is reserved by drawing a 2cm semicircle along the center point. The belly of the qin is dug with a large shovel blade. Attention should be paid not to hurt the reserved part during the digging. The reserved part of longchi and fengzhao is dug down on the slope (small on the top and large on the bottom). The reserved part of the wild yanzu is dug down vertically.

将挖好的琴腹再用小铲刀进行精修。

Refine the dug belly with a small shovel blade.

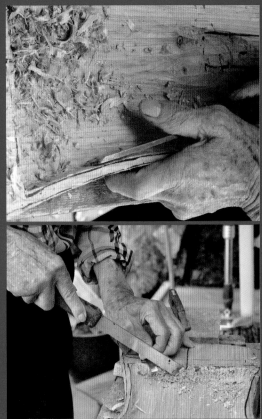

用平铲刀对边墙进行修整，使边墙平整，厚薄一致。

Trim the side wall with a flat shovel blade to make the side wall flat and consistent in thickness.

用刨子刨出龙池、凤沼对应的留实部位，刨的高度为盖上底板后，刚好能容下伸入的手指即可。

Use a plane to dig out the remaining parts corresponding to longchi and fengzhao. The height of the plane is that after the bottom plate is put in place, the place can accommodate a extended finger.

用小刨子修整琴腹，并不停地敲击或用试音弓试音，以达到音色满意为止。

Use a small plane to trim the qin belly, and keep tapping or using a test bow to test the sound until the timbre is satisfactory.

槽腹的制作是否得当直接决定了一张琴音色的好坏，在挖制中多一刀少一刀都会对音色有很大的影响。所以槽腹的制作方法也是历代斫琴师秘不外传的绝技。

　　一般来说古琴的槽腹内侧（七弦）靠边墙的位置厚度为 0.8 至 1 厘米左右，外侧（一弦）靠边墙位置厚度为 1 至 1.5 厘米左右，两边向中间逐渐增厚，近岳山的高音区适当薄一些，可以让声音更通透、洪亮，琴尾适当留厚一些，散音会减小，许多琴三音不和与散音未调整好有很大的关系。当然这些数据与经验不是绝对的，需要根据斫琴材料的不同而进行调整，总体来说面板材料硬度大一些的话就挖制得薄一点，材料偏软的话就得整体留厚一些。

　　古琴的槽腹不是越大越扁声音就越好，要有所控制才能发出动听的声音。舞台使用的演奏琴和自己在安静环境下抚的琴，包括录音使用的琴，要求所表达的音色是不一样的。舞台使用的演奏琴声音要大，观众才可以听得更清楚；安静环境下抚的琴和录音使用的琴，都需要表达出完美的韵味，虽然声音小一些，但在经过录音棚专业声音采集后，那种韵之美才是中国传统古琴艺术的魅力所在（如成公亮先生的"秋籁"琴，音量很小，却声音韵味极佳）。这也是现在常说的"演奏琴"和"文人琴"之分，这种人为区分，与传统琴学思想有密切的关系。传统琴人抚琴是一种自我修养，与他人无关，所以琴音追求的是一种中庸和谐之美，而现在舞台需要的是一种对外表达，"发声"是第一需求，二者之间没有高低、对错之分，都是时代发展的需要。

Whether the ventral groove is properly made is directly related to the timbre of a qin. The timbre varies greatly with one more or one less cut during the digging process. Therefore, the method of making the ventral groove is also a unique skill that the master makers of all ages have kept secret.

Generally speaking, the thickness of the inner side of the ventral groove (seven chords) of the qin is about 0.8-1cm from the sidewall. The thickness of the outer side (one string) to the side wall is about 1-1.5 cm, and the two sides gradually thicken towards the middle. The treble area near the bridge is appropriately thinner, which can make the sound more transparent and loud. The tail of the qin is appropriately thicker to reduce loose sound. The disharmony between the three tones of many qins is closely related to the unadjusted loose sound. Of course, these data and experience are not absolute. They need to be adjusted according to the different materials of the qin. Generally speaking, if the hardness of the panel material is higher, it will be thinner. If the material is soft, it will be thicker as a whole.

It's not the case that the bigger and flatter the ventral groove of the qin, the better the sound. It takes some control measures to produce a beautiful sound. The qin used on the stage and the qin played by oneself in a quiet environment, including the qin used for recording have different timbres. The sound of the qin used on the stage should be loud so that the audience can hear it clearly. Both the qin played in a quiet environment and the qin used for recording need to express perfect tone, though with lower sound. However, after professional sound collection in the recording studio, the beauty of that kind of rhyme is the charm of traditional Chinese Guqin art (such as Mr. Cheng Gongliang's "Qiu Lai" qin, which has a very low volume but excellent tone). This is also the distinction between "qin for performance" and "qin for literati" which are often referred to in modern times. This kind of modern artificial distinction is closely related to the traditional qin theory. The ancient qin players played qin only for self-cultivation and this had nothing to do with others. Therefore, the top priority is to produce beautiful and harmonious sound, which is key to the "golden mean", while the modern stage needs an external expression, and "voice" then comes first. There is neither lowliness nor nobleness, neither right nor wrong, both are but the demand of the times.

第五章 合琴

Chapter V
Assembling

用砂纸将已经挖制好的面板槽腹内部需要签字的地方打磨光滑，以便书写。

Use sandpaper to polish the place that needs to be signed inside the dug-up panel groove belly in order to make it smooth for writing.

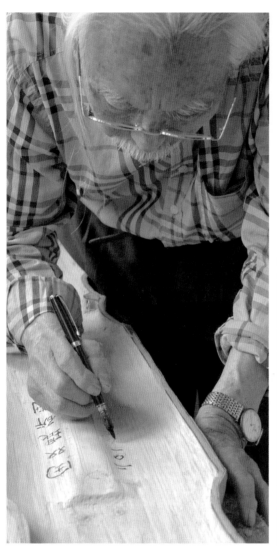
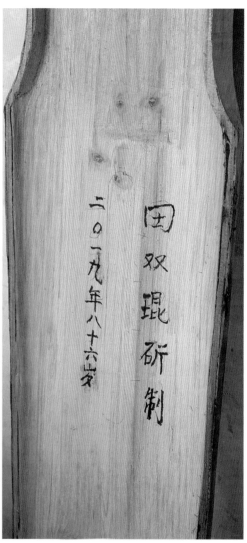

田双琨研制
二〇一九年八十六岁

在琴腹龙池留实两侧签上斫琴师姓名、斫琴时间等信息。

On both sides of the longchi in the belly of the qin, sign information such as when the qin is made and the name of the master maker.

　　用铲刀将雁足留实部位向内铲成斜角，再用平锉将面板边墙向内打磨成斜面，让面底板黏合时不致过于紧密。

Use a shovel blade to shovel the solid part of the wild yanzu inward into an oblique angle, and then use a flat file to polish the surface wall inward into an oblique angle so that the plates will not be too tightly bonded.

对面板、底板各部位进行细致检查，对原来就有的和在斫制中出现的破损、开裂进行修复。

Careful inspection shall be carried out on all parts of the panel and bottom plate to repair the original damages and cracks and those occurred during qin making.

　为了在合琴期间，不让琴出现扭曲、变形等情况，一般与一种叫"矫正板"的平直厚木板一起捆绑。

In order not to distort or deform the qin during the assembling, it is usually bound with a straight thick board called "corrector plate".

合琴前首先检验"矫正板"是否平直、有无变形等情况。

Before assembling the qin, first check whether the "corrector plate" is straight and free from deformation.

依次对面板、底板黏合部位抹胶。牛皮胶、鱼鳔胶等传统生物胶用量可适当大一点，现代高黏性胶可适当少一点，使其既不开脱，又不至于黏合太死。

The bonding parts of the panel bottom plate are plastered with glue in turn, and traditional biological glues such as cowhide glue and swim bladder glue can be appropriately increased. Modern high-viscosity glue can be appropriately reduced, so that the adhesion will neither come unglued nor being too tight.

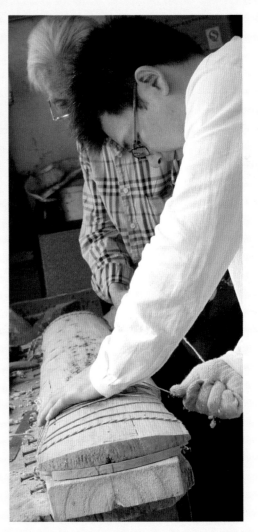

用绳子进行捆绑，观察每一处
面板、底板是否紧密结合。

Bind with ropes and observe
whether the surface and bottom plates
are tightly combined.

捆绑好以后再次进行调整，有未紧密结合处使用木楔子楔入绳子内，直到面板、底板紧密结合为止。

After binding, adjust again. Wedge into the rope with wooden wedges until the surface and bottom plates are tightly connected.

用绳子捆绑好以后，面板、底板如果出现轻微错位，可垫上木方，用榔头适当敲击，进行复位。

After being tied with ropes, if there is a slight dislocation between the surface and the bottom plates, wooden squares can be padded and properly struck with a hammer to reset.

根据选择的合琴用胶种类，一般放置于阴凉干燥处 1 至 3 天即可黏合牢固。

Place the assembled qin in a cool and dry place for 1-3 days for the plates to bind firmly, up to the glue selected and applied.

将已合好的琴体再次打磨、精修，找平边墙。

Polish, refine and level the side walls of the assembled qin body once again.

对内外边墙厚度再次精修、找直等。

The thickness of the inner and outer side walls shall be refined and leveled again.

对底板两侧进行斜面处理，让厚重浑圆的唐代制式琴从视觉上适当减少厚度，以适应现代社会审美标准。

The two sides of the bottom plate are inclined to reduce the thickness of the thick and round Tang-style qin visually to adapt to the aesthetic standards of modern society.

古人说，合琴需挑黄道吉日，这是有一定道理的，对于琴人来说，斫制一张琴是一件十分重要的事情。如果选在不合适的天气、不适合的气温下合琴，对琴的确是有不良影响的。

传统古琴天地柱都很粗，很多历史老琴由于面板过厚，天地柱过粗，导致琴声喑哑、不出音，管平湖先生修这样的琴时一般都会去掉天地柱。我们在制作新琴时也要根据材料的情况、槽腹内部的构造和试音结果而选择是否安装和安装多大尺寸的天地柱。

底板龙池、凤沼的边缘可以镶嵌上红木、竹子等物，起到装饰美观的作用，也可以保护龙池、凤沼边缘不受损伤。如果在合琴前对琴试音时觉得余韵不足的话，在镶嵌的时候可以让底板槽腹的内侧镶饰的红木或者竹子高出 0.1 至 0.2 厘米左右，让声波从龙池、凤沼减缓溢出。

合琴并不是越严越好，用铲刀将雁足留实部位向内铲成斜角，用锉刀将面板边墙打磨成 10 至 15 度左右的斜边，都是为了让面板和底板在黏合的时候不至于过度紧密，有利于古琴的发声。

市场上可供选择的合琴用胶很多，胶干以后可以再打入竹钉，打竹钉的时候，钻的孔眼一定要比竹钉直径稍微大一点儿，眼大竹钉细，留有一定的活动空间，又跑不了。有的斫琴师合琴只采用自攻丝，而且不会拧得太紧，或者拧紧以后再倒回半圈，也是为了让琴出音更好。

有经验的斫琴师在合琴前外部尺寸都会留一定的余量，合琴以后再将琴型外观进行精细修整，以达到外观、音色俱佳的效果。

The ancients said that a qin should be assembled in an auspicious time. This indeed makes sense. For a master maker, qin construction is actually quite a big deal. Choosing an auspicious day can show a good wish. If a qin is made in a day with bad weather and temperature, its quality will suffer.

The heaven and earth pillars of ancient qin are very thick. Many aged qins produce a dull sound or even no sound due to excessively thick panels and heaven and earth pillars. Mr. Guan Pinghu usually removes the heaven and earth pillars when repairing such qins. When making a new-style qin, we should also decide on whether to install the heaven and earth pillars, and if so, what size to install according to the materials selected, the internal structure of the groove belly and the audition results.

The edges of the longchi and fengzhao on the bottom plate can be inlaid with mahogany, bamboo and other objects for decoration use, or to protect the edges of the longchi and fengzhao from damage. If you feel that the rhythm is insufficient during the audition before qin assembling, you can make the mahogany or bamboo inlaid inside the bottom plate about 0.1-0.2 cm higher than the groove belly, so as to slow down the overflow of sound waves from longchi and fengzhao.

It's not the case that the more tightly the qin plates fit, the better the sound. Use a shovel blade to shovel the solid part of the yanzu inward into an oblique angle, and use a file to polish the side wall of the panel into an oblique edge of about 10-15 degrees. The idea here is to prevent the surface and the bottom plates from being bound too tight, which is conducive to the sound of the qin.

There are many kinds of glues for qin assembling available in the market. Such biogums as oxhide gelatin and fish gelatin should be applied more, while the modern high-viscosity chemical glue less so that the adhesion will neither come unglued and nor being too tight. Put bamboo nails after the glue dries. When putting the bamboo nails, the holes drilled must be slightly larger than the bamboo nails. The aim is to leave some room for movement. Some master makers only use Zigong silk to make qin, and they will not screw it too tight. Oftentimes they rewind it half a turn after tightening to make the qin's timbre better.

Experienced master makers will leave a certain margin for the external dimensions before qin assembling. After assembling the qin, the contour will be finely trimmed to achieve the best appearance and timbre.

第六章 装配件

Chapter VI
Assembly Accessories

检查需要装配件的各部位，提前对轸池、护轸等地方进行修整，以便于安装。

Check all accessories to be assembled, and trim the Zhenchi, Huzhen and other places in advance to facilitate installation.

使用本琴形制的模板，画出岳山的确切位置。

Draw the exact location of bridge with the help of the template of the same style with this qin.

用尺子再次测量准确以后，在琴额一侧画出岳山开槽口。

For the sake of accuracy, measure once again with a ruler, and draw the qin forehead side of the bridge's notch.

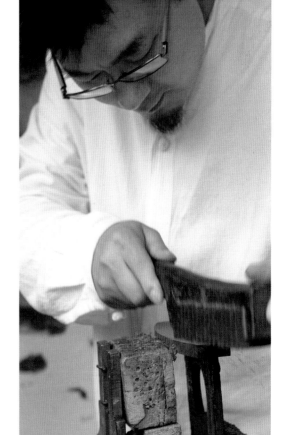

　　使用合适的红木，根据需要安装
岳山部位的琴面宽度和弧度，裁切出
岳山的长度和里外弧度，并进行精细
打磨。

　　With reference to the qin surface
width and radian that to be installed
with the bridge, cut the mahogany to the
length and inner and outer radian of the
bridge, and then finely polish it.

再根据已制作打磨好的岳山的厚度，画出需要开凿的岳山安装槽宽度，用锯锯开两侧。

According to the thickness of the bridge that has been made and polished, draw the width of the installation groove that needs to be excavated, and saw both sides with a handsaw.

用凿子开凿岳山槽，一边凿一边用岳山比对大小和位置高低，使岳山安装槽内的弧度与岳山弧度保持一致。

Dig the bridge groove with a chisel, and compare the size, height, and position with that of the bridge while digging to ensure that the radian of the bridge fits well with that inside the groove.

也可根据情况对岳山进行适当修整。

The bridge can also be properly trimmed as required.

检查岳山是否安装紧密，底部不可留有空隙。

Check whether the bridge is tightly installed to make sure there's no gap at the bottom.

用提前制作好的红木冠角为模板画出
冠角安装槽的位置和形状。

Draw the position where the crown angle
is to be installed with the mahogany crown
angle made in advance.

用平铲沿画好的线铲出冠角的安装槽，同时铲好龙
龈安装的位置。

Use a flat shovel to shovel out the installation groove
required for the crown angle along the drawn line, as well as
the installation position of dragon's gums.

调整龙龈大小，修整冠角和龙龈的黏合部位，让二者紧密贴合。

Adjust the size of the dragon's gums and trim the bonding parts of the crown angle and the dragon's gums so that the two fit tightly.

检查冠角和龙龈与琴体是否紧密贴合。

Check whether the crown angle and dragon's gums fit closely with the qin body.

画出托尾与龈托位置，为避免安装偏位，需提
前找好中心位置。

Draw the position of the tray tail and gums tray.
Find the central position in advance to prenvent install
deviation.

用平铲铲出托尾和龈托安装的位置。

Shovel out the installation position of the tray tail and gums tray with a flat shovel.

检查托尾和龈托与琴体是否平整、紧密地贴合在一起。

Check whether the tray tail and gums tray are flat and tightly attached to the qin body.

　　将所有准备好的配件和要安装的位置用胶抹匀，用钉子、
夹子、绳子等固定。

　　Apply glue on all prepared accessories and positions to be
installed evenly and fix them with nails, clips, ropes, etc.

再次对黏合配件的位置进行调整，使其精确到位，放置阴凉干燥处待干。

Adjust the position of the bonded accessories once again for the sake of accuracy, place them in a cool and dry place for drying.

岳山和龙龈的施工十分重要，它们与两边贴得紧不紧密、下边有没有空隙，对声音的传导都有很大的影响，是继槽腹之后另外一个对声音有直接影响的元素，这种影响可以灵活运用，与已制作好的槽腹相互配合，进行松紧调配，可取得良好的音色效果。

岳山在安装的时候，一般只黏中间位置，如果因为音色处理原因，让岳山槽略微宽松一点的话，用皮胶进行全面黏合即可。

红木配件既有保护古琴重要部位的作用，又是一种十分重要的装饰，所有配件前期的制作和安装到位后的整体修型、雕刻都需要精益求精，以达到一种形神兼备的完美状态。

The construction of the bridge and dragon's gums is of crucial importance. Whether they fit tightly with the two sides or whether there is a gap at the bottom have a great influence on the transmission of sound. Aside from the groove belly, they are another two elements that have a direct influence on the sound of qin. If flexibly adjusted, they can work closely with the already made groove belly for a better timbre.

When installing the bridge, applying glue on its middle part is enough. However, if a little looser bridge groove is needed for timbre processing purposes, the whole area can be glued with leather glue.

Mahogany accessories, besides protecting the qin from potential damages, serve as a key decoration for the qin. Throughout the entire process, from early production of all accessories, the overall patching to carving after installation, the master makers must seek for perfection to make the qin of both excellent timbre and exquisite shape.

第七章　裱葛布

Chapter VII
Ko-hemp Cloth
Mounting

先将琴体小型凹陷部分用粗灰胎进行修补，以便裱的
葛布可以与琴体紧密贴合。

Repaire the small sunken parts of the qin body with coarse
gray tire, so that the mounted ko-hemp cloth can fit closely with
the qin body.

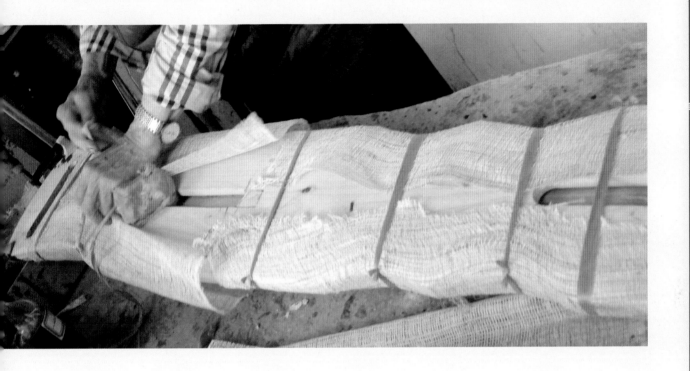

将葛布裁成合适大小，用细绳子捆于琴体上，然后向葛布喷水，湿润以后调整项、腰以及转角处，使葛布与琴体紧密结合。

Cut the ko-hemp cloth into pieces of appropriate size, tie them to the qin body with a thin rope, then spray water on them, adjust the qin neck, waist and corners after the cloth get wet, so that the ko-hemp cloth is closely combined with the qin body.

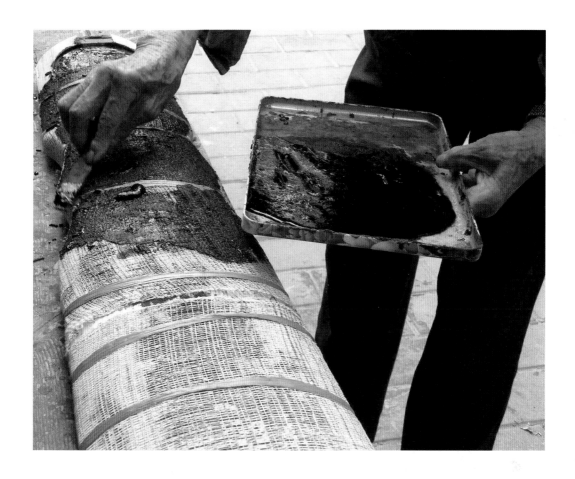

　　将生漆、鹿角霜细灰（100目）调和成灰浆。调和好以后静置半个小时左右，待氧化变黑，再次加入适量生漆，调和到刮刀铲起时可以往下滴落，同时又不会太稀为好。等贴敷好的葛布半干以后，用刮刀薄而均匀地刮到贴敷于琴面的葛布上，等待干燥到不沾手时，去除细绳，再将细绳部位补刮上灰浆。

　　Mix the raw lacquer and fine ashes of cornu cervi degelatinatum (100 mesh) into mortar. After blending, let it sit for about half an hour, add a proper amount of raw lacquer again after it turns black due to oxidation, blend the mortar to the extent that it can drip down from the scraper blade yet not too thin. Apply the mortar with a scraper blade on the adjusted and pasted ko-hemp cloth after it is semi-dry. Wait until it is dried to the point where it does not stick to the hand, remove the rope and then paste the mortar to the position of the rope.

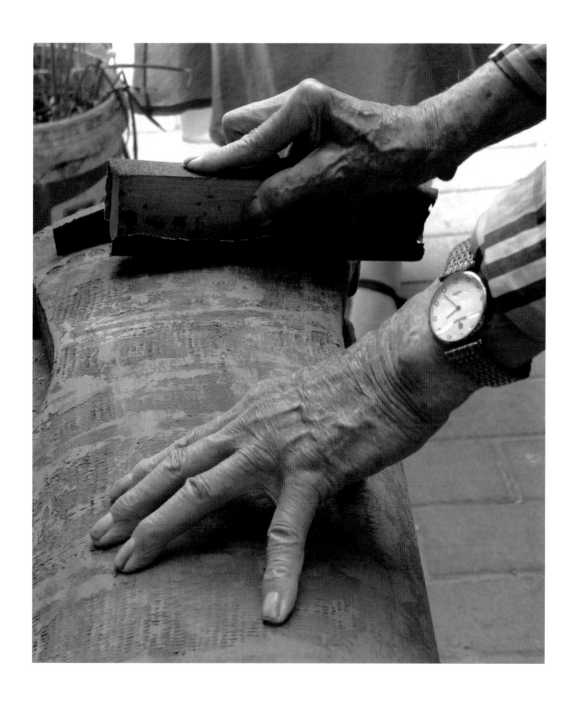

　　放置 7 至 15 天，待灰浆干燥后，仔细检查各部位的葛布是否粘贴紧密，如有粘贴不实或翘起处，用美工刀剔除干净，再用粗砂纸对表面进行大致打磨，不宜打磨过重，以免伤及葛布。

Leave it for 7-15 days. After the mortar is dried, check carefully whether the ko-hemp cloth is tightly pasted in each part. If there is any part not pasted tightly, remove it with a box cutter, and then polish the surface with coarse sandpaper. Remember not to polish it too heavily, otherwise damages may be made to the ko-hemp cloth.

琴在施工中裱的布称之为"葛布"，它是由纯苎麻编织而成，过去也有不用编织的葛布，用纯苎麻一缕一缕地直接裱糊到琴上。

古代的琴有些裱布，有些不裱，这与南北气候差异有关。南方通常温润潮湿，琴体不易开裂，北方空气干燥，早晚温差大，再加上现在北方冬季供暖，琴体很容易干裂、变形，裱葛布是解决这一难题的最好办法。唐宋时期大部分琴都裱有葛布，流传到现在保存得也都比较完好，这与那个时代弹琴、斫琴的人多在北方有关。南宋至明清时期南方文风渐盛，这时候弹琴、斫琴者多出自南方，裱葛布的就比较少。也有的琴裱麻纸或宣纸，但都出现了严重的脱落，经不起时间的考验。

琴体裱葛布会对琴的音色有极大的影响，需要在槽腹制作时提前考虑进去，槽腹与裱布搭配得当，会收到意想不到的良好效果。一般来说葛布轻薄、细密一点，用原生漆直接裱到琴体上的，多出皮鼓样声音；葛布厚重、粗而稀疏一点，用调配好的细灰浆裱糊到琴体上，多出苍古的声音。

The cloth mounted on the qin body during qin making is referred to as "ko-hemp cloth". It is woven from pure ramie fiber. In the past, some master makers just pasted the pure ramie fiber directly on the qin body without weaving it.

In history, some qins were mounted yet some not, which has something to do with the climate difference between the north and south. The warm and humid climate in the south prevents the qin body from cracking, while the day air, coupled with the huge temperature difference between mornings and evenings and modern heating system in the north cause the qin body to crack and deform easily, and ko-hemp cloth mounting is the best solution for this problem. Most of the qins in the Tang and Song Dynasties were mounted with ko-hemp cloth, which could be well preserved even in modern times. This is because most of the master makers and qin players then lived in the north. From the Southern Song Dynasty to the Ming and Qing Dynasties, literature and arts flourished in the south, and a large number of master makers and qin players emerged at this region. Consequently, the majority of qins made at this time period are not mounted. There are also qins mounted with hemp paper and rice paper, all fail to stand the test of times and fall off from the qin body seriously.

The mounting of ko-hemp cloth on the qin body has a tremendous impact on the timbre, which thus should be taken into account in advance when making the groove belly. A proper blending of the groove belly and the mounting cloth will achieve unexpected good results. Generally speaking, the qin will produce drum-like sounds when light, thin and finer ko-hemp cloth is mounted directly on the qin body using raw lacquer, while making more ancient sounds when mounted with thick, coarse and sparse ko-hemp cloth using thinner fine mortar.

第八章　刮灰胎

Chapter VIII
Scraping Lacquer
Cement

刮灰胎用的泥子一般是使用鹿角霜和生漆
调和而成。

While scraping lacquer cement, the putty
used is usually made by blending cornu cervi
degelatinatum and raw lacquer.

　　生漆与鹿角霜搅拌均匀后，静置半个小时左右，待鹿角霜充分吸收生漆，再次进行搅拌，根据搅拌的手感感受灰胎泥子的软硬度，如果过硬则再次加入生漆进行调和，过干、过硬会不便于施工，过稀、过软则不利于干燥且干燥中容易起皱。

　　Stir evenly the cornu cervi degelatinatum and raw lacquer, let the resulting mixture stand for about half an hour for the former to fully absorb the latter. Stir again, and feel the hardness of the putty according to the hand feeling. Add additional raw lacquer again for blending if it is too hard. Because too dry or hard putty will make the qin construction process more difficult, while too thin and too soft putty is not conducive to drying and may result in wrinkling.

　　把泥子用刮刀均匀横刮到琴体上，琴面要求中间稍厚，两边稍薄，琴底则厚薄均匀。

　　Scrape the putty evenly and horizontally on the qin body with a scraper blade, with that on the middle part of the qin surface being slightly thicker than on both sides. On the bottom of qin, however, make sure to scrape the putty with exactly the same thickness.

横刮一段泥子以后，再使用刮刀轻轻进行竖刮，以便灰胎更平整、光滑。

Scrape vertically using a scraper blade after scraping horizontally for a while to make the lacquer cement smoother and more glossy.

注意琴体的转角处，需要用刮刀将泥子精心填实、刮平，找出 90 度直角。

Pay much attention when scraping the corner of the qin body. Make sure the putty is carefully filled and scraped with a scraper blade and find the 90 degree right angle.

护轸位置在第一遍刮粗泥子的时候就得耐心地塑造出大致的形状，然后经过多次打磨和刮泥子塑形，才能达到完美效果。

Make the rough shape of Huzhen when scraping the putty for the first time, and then polish and scrape it repeatedly for perfection.

用刮刀将少量泥子均匀地刮于琴体
的内外边墙。

Use a scraper blade to evenly scrape a
small amount of putty on the inner and outer
side walls of the qin body.

再用刮刀从侧面轻轻找出 90 度直角，注意不要将面底已
刮好的灰胎翻起。

Then find a 90 degree right angle from the side with the same
scraper blade. Be careful not to turn up the lacquer cement that has
been scraped at both the surface and bottom of the qin.

用刮刀尖刮出琴头的凤舌，凤舌处需用泥子少量多次刮塑，直到令人满意为止。

Scrape out the phoenix tongue of the qin head with the tip of the scraper blade. Scrape putty on the phoenix tongue for several times until it looks nice.

用泥子刮平琴尾，注意龙龈与
龈托之间的凹入部分，需要平整光
滑，转角处宜处理为圆润的钝角。

Apply putty on the tail of qin to
make it smooth. Make sure that the
concave part between the dragon's
gums and the gums tray is glossy and
smooth. The corners should be polished
to round obtuse angles.

古琴的灰胎历史上多用生漆调和鹿角霜制成，因鹿角霜有天然微孔，软硬适中，经过一两千年的实际使用总结出的经验是，这样出音效果和音色是最好的。又因鹿在中华传统文化中象征长春、吉祥之意，故鹿角霜作为斫琴主要原材料被普遍使用，奉为上品。灰胎也有用八宝灰、孜子灰（打磨翡翠沉淀下来的灰浆）、瓦灰、瓷粉、菟丝子灰等制作的。

生漆在我国的使用历史十分悠久，可以追溯到8000年前。生漆是自然生长的漆树的树汁，它含有的漆酶成分在液态时容易引起过敏反应，干燥以后则环保无毒。现代也有用合成漆、化学漆的，特别是一些南方地区大批量生产的低端厂琴，均使用合成漆和化学漆，音质效果不佳，而且合成漆、化学漆含有苯、铅等大量有害物质，长期放置在室内和坐在其跟前弹奏，对身体有极大的伤害。

灰胎的厚薄与古琴的槽腹大小、面板挖制的厚薄成反比，它们相互结合，才能达到古琴音色传统的中庸和谐之美。

上灰胎的时候一般来说先上粗灰，待干燥以后用粗砂纸打磨平整，再上中灰，用中号砂纸打磨，再上细灰，用细砂纸打磨。刮灰胎的施工要求一遍比一遍平整精确，粗灰阶段就要求基本达到平整，特别是在弦路部位，不可留有过大的凹陷或缺损，否则后期会由于灰胎粗细收缩率不同而出现煞音。

粗、中、细三遍灰胎法是一种传统的常规方法，在实际应用中，可以薄而多遍地刮灰胎、打磨，出来的声音层次感、颗粒感会更强，也可以粗细顺序颠倒，或者配合部分瓷粉、瓦灰和不同种类的生漆等交叉使用，可以获得"金石""皮鼓"等意想不到的音色效果。

Historically, Guqin's lacquer cement is made by blending raw lacquer and cornu cervi degelatinatum. Because cornu cervi degelatinatum has natural micropores and a moderate hardness. Actual experience after 1,000 to 2,000 years of use shows that a qin will have the best sound and timbre if it is scraped with such kind of lacquer cement. In traditional Chinese culture, deer is believed to contain auspicious messages with longevity and luck. The cornu cervi degelatinatum, therefore, is widely used as a main raw material for lacquer cement in making qin. Other commonly seen materials for lacquer cement including Babao ash, Zizi ash (mortar precipitated by polishing jadeite), tile ash, porcelain powder, dodder ash, etc.

The use of raw lacquer on physical objects in China can be traced back to 8,000 years ago. Raw lacquer is the sap of naturally growing lacquer trees. The laccase component contained in raw lacquer can easily cause allergic reactions in liquid state, yet is environmentally friendly and non-toxic after drying. Some synthetic lacquers and chemical lacquers are also used in modern times, especially in qins made by some low-end factories in large quantities in southern China. Such qins often have poor timber. Moreover, synthetic lacquers and chemical lacquers contain a large number of harmful substances such as benzene and lead, which, when applied to a qin and placed indoors for a long time, will cause great harm to human body.

The thickness of the lacquer cement is inversely proportional to the size of the groove belly and the surface panel of qin. Only if that is ensured, the qin will produce harmonious and euphonic timbre.

Generally speaking, when applying the lacquer cement, coarse cement is applied first. After it gets dry, polish with the coarse sandpaper till it becomes glossy. Then, medium cement is applied, polished with medium sandpaper, and finally fine cement, and polished with fine sandpaper. In scraping lacquer cement, the three steps are becoming more and more demanding. In the coarse ash stage, a basic requirement is smoothness, especially in the chord part, no excessive depression or defect should be left, otherwise the shrinkage rate of ashes may result in bad sound.

The three-step (coarse, medium and fine cements) scraping method is a conventional one. In practice, though, scraping and polishing can be done repeatedly, with a thin layer of lacquer cement being applied each time. A qin made in this way will produce layered sound. The order of the three steps can also be reversed. Some unexpected timbres such as "stone" and "leather drum" may be achieved by using a mixture of materials like porcelain cement, tile ash and lacquers of different types.

第九章　磨灰胎

Chapter IX
Polishing Lacquer
Cement

首先打磨琴面位置，争取第一遍粗灰胎弧度就舒适、平整。

First of all, polish the qin surface to make the radian of the coarse cement flat and smooth.

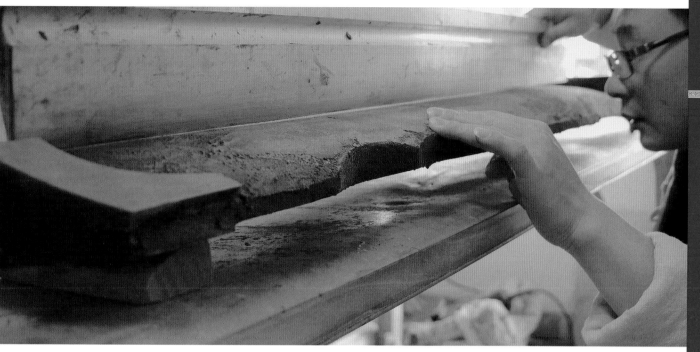

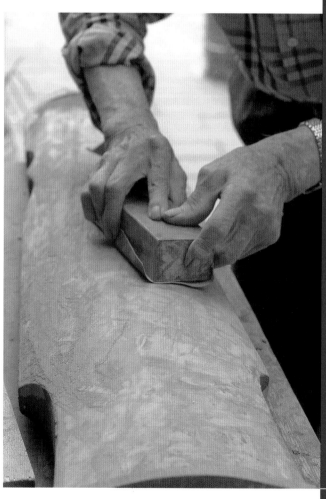

　　用靠尺反复对比，特别是弦路位置部分，需要平直到位，并磨出适当的塌腰，才不会出现煞音、打板等问题。

　　Use a guiding rule for repeated comparison and reference, and make sure the position of the chord road is straight. Grind out an appropriate waist collapse to prevent problems such as bad sound and board beating.

　　打磨护轸位置一定要双侧对比进行，注意塑造护轸的形状。

　　Take both sides into account when polishing the position of Huzhen, then sculpt the shape of Huzhen.

　　打磨岳山、冠角附近的灰胎时，一定注意不要磨损
岳山、冠角。

　　Be careful not to cause damages to the bridge and
crown angle when polishing the lacquer cement near these
two parts.

　　打磨古琴边墙时，要求平整、顺直，可用靠尺进行参考，也可提前画出边线，根据边线进行打磨。

　　Make sure the side walls of the Guqin are flat and straight while polishing them. A guiding rule may be helpful, or you can draw sidelines in advance and polish according to the sidelines.

琴项、腰部位的打磨需要时刻关注弧度的合理性和双侧
的一致性。

When polishing the qin neck and waist, make sure the radians
are appropriate, and both sides consistent.

灰胎在干燥过程中会出现收缩，用漆量越大、灰越细，收缩率越高。所以从刮灰胎的施工开始必须每次基本平整，不能留有高低不平或者凹陷，如果施工中出现高低不平和凹陷的情况，必须使用砂纸打磨平整，才可以进行下一次的刮灰胎施工，尤其是弦路部分要特别注意。

粗灰打磨的时候可以用靠尺进行参考，力求平整。从中、细灰开始，打磨的时候要交替使用靠尺和试音弓，最后以试音弓在弦路位置试音，以无打板、煞音，其他地方平整光滑为准。

The lacquer cement will shrink when drying. The more lacquer cement you use and the finer the cement, the greater the shrinking percentage. Therefore, smoothness becomes a mast for each and every step since scraping lacquer cement. If some parts are left rugged during the construction, use the sandpaper to polish and level it off before the next step, especially at the position of the chord road.

When polishing with coarse cement, a guiding rule can be used for reference to make the surface smooth. Use the guiding rule and audition bow alternately during polishing the medium and fine cements. Make sure that when testing the audition bow at the chord road position, it will not beat the board, and that the sound produced is good, and the rest parts are smooth.

第十章 补灰胎

Chapter X
Lacquer Cement
Mending

用 120 至 200 目鹿角霜与生漆调和为灰浆，静置半小时左右，待氧化变黑，再次加入适量生漆，调和到铲起时会往下滴，同时又不会太稀为好，寻找已磨好灰胎的沙眼以及经过粗砂纸打磨的划痕等进行修补。

Mix 120-200 mesh of cornu cervi degelatinatum with raw lacquer to mortar, let the mixture stand for about half an hour, add a proper amount of raw lacquer again after it turns black due to oxidation, blend the mortar to the extent that it can drip down from the scraper blade yet not too thin. Then paste the mortar on the sand holes formed after applying the lacquer cement and scratches caused by polishing with coarse sandpaper.

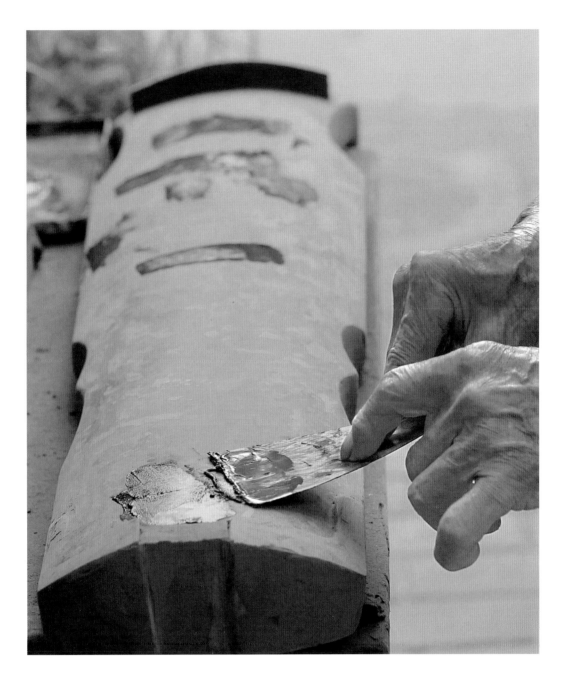

　　特别注意要对岳山、冠角、龙龈、护轸等部位与灰胎连接处和各部位的转角处进行细致的修补，如转角处有较大的缺损，第一遍修补以后等几分钟，让灰胎吸收部分灰浆水分，稍干以后再次修补。

　　Pay special attention to the connections of the bridge, crown angle, dragon's gums, Huzhen and other parts with the lacquer cement, as well as the corners of each part when repairing. If there are large defects at corners, wait a few minutes after the first repair for the lacquer cement to absorb part of the mortar moisture, and then repair for a second time.

当琴体灰胎无明显的沙眼、缺损、凹陷时，再用适
量细灰浆通体刮遍，此步骤又称之为"闭毛孔"。

When there are no obvious sand holes, defects or hollows
in the lacquer cement of the qin body, apply a proper amount
of fine mortar on the whole body. This step is also referred to
as "pore shutting".

待修补的灰浆干燥以后，用细砂纸进
行细致的打磨。

Polish the mortar after it is dried carefully
with fine sandpaper.

补灰胎与闭毛孔的工作十分重要，因为接下来琴面髹漆使用的原生漆、黑推光漆、精炼漆等无良好的填充性，琴体灰胎上遗留的沙眼、划痕，转角处微量的缺损，上徽位时（上徽位也可以在本次工序之前，便于补灰胎的工作一次完成）灰浆收缩留下的凹陷等必须由本次工作修补完成。

琴体灰胎的沙眼、缺损、凹陷部位用灰浆使用小刮刀逐一刮填平整，干透以后用细砂纸精细打磨，待无明显的沙眼、缺损、凹陷时，再用适量细灰浆（生漆调和 200 目鹿角霜而成）通体刮遍。也可以用棉纱先蘸少量生漆，再蘸少许滑石粉或 200 目以上鹿角霜在灰胎上进行揉抹，这在漆艺中又称之为"揉粉"，主要作用是填补灰胎的粗大毛孔。等待彻底干燥以后，再用 600 至 800 目细砂纸通体打磨光滑即可。

Pore shutting is no less important than acquer cement mending. Because the sand holes, scratches and defects at corners left on the qin body due to the poor fillibility of raw lacquer, black polished lacquer, refined lacquer, etc. during lacquer cement pasting, and the hollows formed by mortar shrinking amid amid the work for the ensemble positions (the ensemble positions can be done before this step so that lacquer cement mending can be completed at once) must be mended at this step.

Scrape and level off the sand holes, defects and hollow parts of the lacquer cement of the qin body with mortar using a small scraper blade. Polish it with fine sandpaper after drying until there are no obvious sand holes, defects and hollow parts. Scrape the whole body with an appropriate amount of fine mortar (blended from raw lacquer and 200-mesh cornu cervi degelatinatum). You can also use cotton yarn dipped in a small amount of raw lacquer, then stick a little talcum powder or cornu cervi degelatinatum of more than 200 mesh to rub on the lacquer cement, which is also called "rubbing powder" in lacquer art, and its main function is to fill the thick pores of the lacquer cement. Polish the whole body after thorough drying using 600-800 mesh fine sandpaper.

第十一章　裝琴徽

Chapter XI
Emblem
Mounting

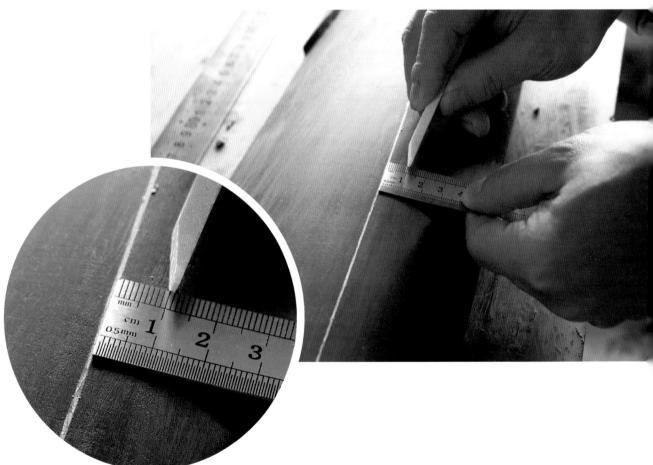

　　用长尺画出一弦的弦路位置，在靠近岳山、龙龈的地方分别向外标出 1 厘米的点，两点相连画出一条距一弦弦路位置 1 厘米的线作为琴徽的安装位置。

　　Draw the path of string-1 with a long ruler, mark the 1cm points outward near the bridge and dragon's gums, respectively, and connect the two points to draw a parallel line 1cm apart as the installation position of the emblem.

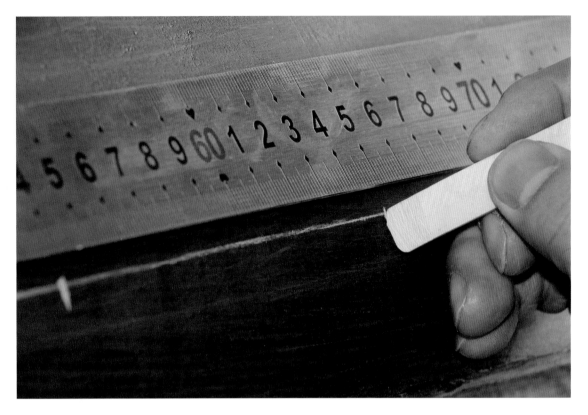

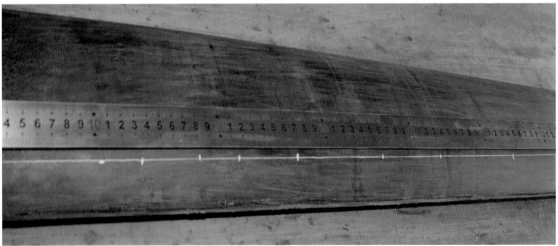

根据古琴有效弦长（岳山与龙龈之间的长度）测算出
各个徽位之间的距离，用白笔画好标记。

Measure the distance between each emblem according to
the effective string length of the Guqin (the length between the
bridge and dragon's gums), and mark with a white pen.

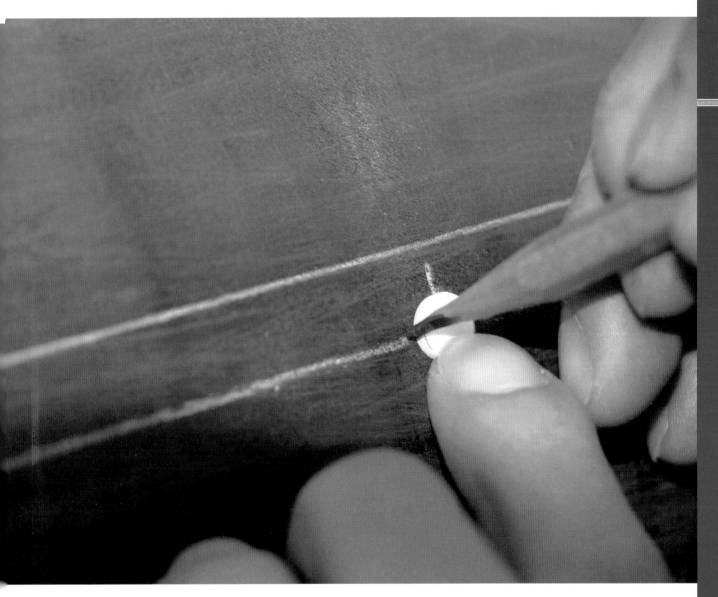

按照各琴徽大小用细尖铅笔画出需要开孔的位置，一定要准确到位。

Draw the hole position for installing the emblem with a sharp pencil based on the size of each emblem, and make sure the installation position is accurate.

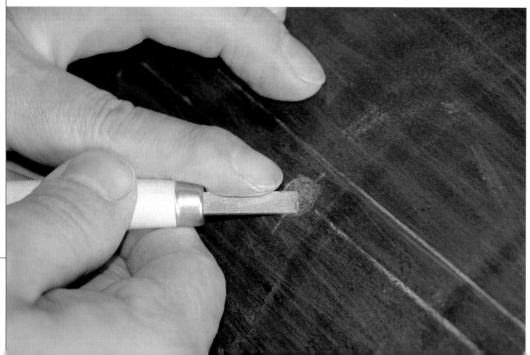

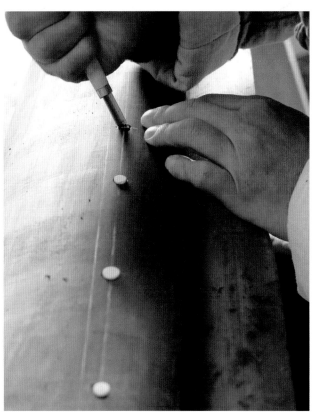

用刻刀按照画好的位置逐一开好徽位安装孔，将提前制作好的琴徽放入安装孔测试好孔的大小深浅。

Use a carving knife to drill holes for the installation of emblems one by one on the pre-drawn positions, and put the pre-made qin emblems into the holes to test their size and depth.

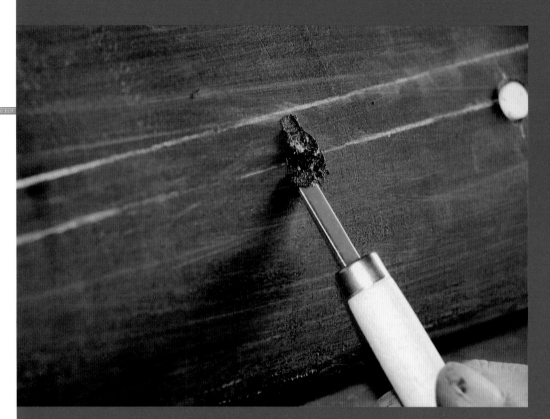

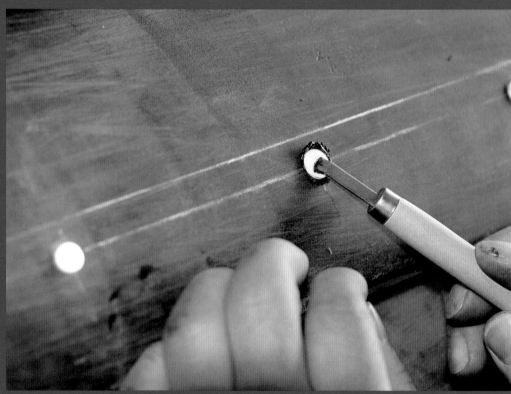

　　将细灰胎填入开好的安装孔，用平刻刀把琴徽压入孔内，回收抹平溢出的多余灰胎，等待干燥。

　　Fill the fine lacquer cement into the holes, press the qin emblems into the holes with a flat carving knife, reclaim and smooth the overflowing lacquer cement, and wait for it to dry.

安装琴徽的灰胎干燥以后，使用砂纸打磨平整。

After the lacquer cement used in installing the qin emblem is dried, polish and level it with sandpaper.

再次测量琴徽安装位置是
否准确。

Measure again to determine
whether the emblems are installed
in the right position.

用细砂纸精细水磨后
等待下一步髹面漆。

Sand down with fine
waterproof abrasive paper
before painting the topcoat.

徽位是根据三分损益的原则来确定，按照有效弦长的二分法、三分法、五分法、六分法、八分法而确定，具体徽位计算方法如下：

二等分：1/2 处·七徽；

三等分：1/3 处·五徽；2/3 处·九徽；

四等分：1/4 处·四徽；2/4 处·七徽；3/4 处·十徽；

五等分：1/5 处·三徽；2/5 处·六徽；3/5 处·八徽；

4/5 处·十一徽；

六等分：1/6 处·二徽；2/6 处·五徽；3/6 处·七徽；

4/6 处·九徽；5/6 处·十二徽；

八等分：1/8 处·一徽；2/8 处·四徽；4/8 处·七徽；

6/8 处·十徽；7/8 处·十三徽。

The emblem position is delineated by the principle of three scale fall and rise, and is usually determined by dividing the effective string length into 2, 3, 5, 6 and 8 equal parts. The specific calculation method for emblem position is as follows:

2 equal parts: 1/2 of the way up the string • Seven emblems;

3 equal parts: 1/3 of the way up the string • Five emblems; 2/3 of the way up the string • Nine emblems;

4 equal parts: 1/4 of the way up the string • Four emblems; 2/4 of the way up the string • Seven emblems; 3/4 of the way up the string • Ten emblems;

5 equal parts: 1/5 of the way up the string • Three emblems; 2/5 of the way up the string • Six emblems; 3/5 of the way up the string • Eight emblems;4/5 of the way up the string • Eleven emblems;

6 equal parts: 1/6 of the way up the string • Two emblems; 2/6 of the way up the string • Five emblems; 3/6 of the way up the string • Seven emblems;4/6 of the way up the string • Nine emblems; 5/6 of the way up the string • Twelve emblems;

8 equal parts: 1/8 of the way up the string • One emblem; 2/8 of the way up the string • Four emblems; 4/8 of the way up the string • Seven emblems;6/8 of the way up the string • Ten emblems; 7/8 of the way up the string • Thirteen emblems.

常用古琴徽位尺寸表
(Dimension Table of the Emblem Position of Common Guqin)

徽位 Emblem Position / 尺寸 Size (cm)	1	2	3	4	5	6	7	8	9	10	11	12	13
108.4	13.55	18.07	21.68	27.10	36.13	43.36	54.2	65.0	72.3	81.3	86.7	90.3	94.9
108.5	13.56	18.08	21.70	27.13	36.17	43.4	54.25	65.1	72.3	81.4	86.8	90.4	94.9
108.6	13.58	18.10	21.72	27.15	36.20	43.44	54.3	65.2	72.4	81.5	86.9	90.5	95.0
108.7	13.59	18.12	21.74	27.18	36.23	43.48	54.35	65.2	72.5	81.5	87.0	90.6	95.1
108.8	13.60	18.13	21.76	27.20	36.27	43.52	54.4	65.3	72.5	81.6	87.0	90.7	95.2
108.9	13.61	18.15	21.78	27.23	36.30	43.56	54.45	65.3	72.6	81.7	87.1	90.8	95.3
109.0	13.63	18.17	21.80	27.25	36.33	43.6	54.5	65.4	72.7	81.8	87.2	90.8	95.4
109.1	13.64	18.18	21.82	27.28	36.37	43.64	54.55	65.5	72.7	81.8	87.3	90.9	95.5
109.2	13.65	18.20	21.84	27.30	36.40	43.68	54.6	65.5	72.8	81.9	87.4	91.0	95.6
109.3	13.66	18.22	21.86	27.33	36.43	43.72	54.65	65.6	72.9	82.0	87.4	91.1	95.6
109.4	13.68	18.23	21.88	27.35	36.47	43.76	54.7	65.6	72.9	82.1	87.5	91.2	95.7
109.5	13.69	18.25	21.90	27.38	36.50	43.8	54.75	65.7	73.0	82.1	87.6	91.3	95.8
109.6	13.70	18.27	21.92	27.40	36.53	43.84	54.8	65.8	73.1	82.2	87.7	91.3	95.9
109.7	13.71	18.28	21.94	27.43	36.57	43.88	54.85	65.8	73.1	82.3	87.8	91.4	96.0
109.8	13.73	18.30	21.96	27.45	36.60	43.92	54.9	65.9	73.2	82.4	87.8	91.5	96.1
109.9	13.74	18.32	21.98	27.48	36.63	43.96	54.95	65.9	73.3	82.4	87.9	91.6	96.2
110.0	13.75	18.33	22.00	27.50	36.67	44	55	66.0	73.3	82.5	88.0	91.7	96.3
110.1	13.76	18.35	22.02	27.53	36.70	44.04	55.05	66.1	73.4	82.6	88.1	91.8	96.3
110.2	13.78	18.37	22.04	27.55	36.73	44.08	55.1	66.1	73.5	82.7	88.2	91.8	96.4
110.3	13.79	18.38	22.06	27.58	36.77	44.12	55.15	66.2	73.5	82.7	88.2	91.9	96.5

110.4	13.80	18.40	22.08	27.60	36.80	44.16	55.2	66.2	73.6	82.8	88.3	92.0	96.6
110.5	13.81	18.42	22.10	27.63	36.83	44.2	55.25	66.3	73.7	82.9	88.4	92.1	96.7
110.6	13.83	18.43	22.12	27.65	36.87	44.24	55.3	66.4	73.7	83.0	88.5	92.2	96.8
110.7	13.84	18.45	22.14	27.68	36.90	44.28	55.35	66.4	73.8	83.0	88.6	92.3	96.9
110.8	13.85	18.47	22.16	27.70	36.93	44.32	55.4	66.5	73.9	83.1	88.6	92.3	97.0
110.9	13.86	18.48	22.18	27.73	36.97	44.36	55.45	66.5	73.9	83.2	88.7	92.4	97.0
111.0	13.88	18.50	22.20	27.75	37.00	44.4	55.5	66.6	74.0	83.3	88.8	92.5	97.1
111.1	13.89	18.52	22.22	27.78	37.03	44.44	55.55	66.7	74.1	83.3	88.9	92.6	97.2
111.2	13.90	18.53	22.24	27.80	37.07	44.48	55.6	66.7	74.1	83.4	89.0	92.7	97.3
111.3	13.91	18.55	22.26	27.83	37.10	44.52	55.65	66.8	74.2	83.5	89.0	92.8	97.4
111.4	13.93	18.57	22.28	27.85	37.13	44.56	55.7	66.8	74.3	83.6	89.1	92.8	97.5
111.5	13.94	18.58	22.30	27.88	37.17	44.6	55.75	66.9	74.3	83.6	89.2	92.9	97.6
111.6	13.95	18.60	22.32	27.90	37.20	44.64	55.8	67.0	74.4	83.7	89.3	93.0	97.7
111.7	13.96	18.62	22.34	27.93	37.23	44.68	55.85	67.0	74.5	83.8	89.4	93.1	97.7
111.8	13.98	18.63	22.36	27.95	37.27	44.72	55.9	67.1	74.5	83.9	89.4	93.2	97.8
111.9	13.99	18.65	22.38	27.98	37.30	44.76	55.95	67.1	74.6	83.9	89.5	93.3	97.9
112.0	14.00	18.67	22.40	28.00	37.33	44.8	56	67.2	74.7	84.0	89.6	93.3	98.0
112.1	14.01	18.68	22.42	28.03	37.37	44.84	56.05	67.3	74.7	84.1	89.7	93.4	98.1
112.2	14.03	18.70	22.44	28.05	37.40	44.88	56.1	67.3	74.8	84.2	89.8	93.5	98.2
112.3	14.04	18.72	22.46	28.08	37.43	44.92	56.15	67.4	74.9	84.2	89.8	93.6	98.3
112.4	14.05	18.73	22.48	28.10	37.47	44.96	56.2	67.4	74.9	84.3	89.9	93.7	98.4
112.5	14.06	18.75	22.50	28.13	37.50	45	56.25	67.5	75.0	84.4	90.0	93.8	98.4
112.6	14.08	18.77	22.52	28.15	37.53	45.04	56.3	67.6	75.1	84.5	90.1	93.8	98.5
112.7	14.09	18.78	22.54	28.18	37.57	45.08	56.35	67.6	75.1	84.5	90.2	93.9	98.6

续表 Continued

112.8	14.10	18.80	22.56	28.20	37.60	45.12	56.4	67.7	75.2	84.6	90.2	94.0	98.7
112.9	14.11	18.82	22.58	28.23	37.63	45.16	56.45	67.7	75.3	84.7	90.3	94.1	98.8
113.0	14.13	18.83	22.60	28.25	37.67	45.2	56.5	67.8	75.3	84.8	90.4	94.2	98.9
113.1	14.14	18.85	22.62	28.28	37.70	45.24	56.55	67.9	75.4	84.8	90.5	94.3	99.0
113.2	14.15	18.87	22.64	28.30	37.73	45.28	56.6	67.9	75.5	84.9	90.6	94.3	99.1
113.3	14.16	18.88	22.66	28.33	37.77	45.32	56.65	68.0	75.5	85.0	90.6	94.4	99.1
113.4	14.18	18.90	22.68	28.35	37.80	45.36	56.7	68.0	75.6	85.1	90.7	94.5	99.2
113.5	14.19	18.92	22.70	28.38	37.83	45.4	56.75	68.1	75.7	85.1	90.8	94.6	99.3
113.6	14.20	18.93	22.72	28.40	37.87	45.44	56.8	68.2	75.7	85.2	90.9	94.7	99.4
113.7	14.21	18.95	22.74	28.43	37.90	45.48	56.85	68.2	75.8	85.3	91.0	94.8	99.5
113.8	14.23	18.97	22.76	28.45	37.93	45.52	56.9	68.3	75.9	85.4	91.0	94.8	99.6
113.9	14.24	18.98	22.78	28.48	37.97	45.56	56.95	68.3	75.9	85.4	91.1	94.9	99.7
114.0	14.25	19.00	22.80	28.50	38.00	45.6	57	68.4	76.0	85.5	91.2	95.0	99.8
114.1	14.26	19.02	22.82	28.53	38.03	45.64	57.05	68.5	76.1	85.6	91.3	95.1	99.8
114.2	14.28	19.03	22.84	28.55	38.07	45.68	57.1	68.5	76.1	85.7	91.4	95.2	99.9
114.3	14.29	19.05	22.86	28.58	38.10	45.72	57.15	68.6	76.2	85.7	91.4	95.3	100.0
114.4	14.30	19.07	22.88	28.60	38.13	45.76	57.2	68.6	76.3	85.8	91.5	95.3	100.1
114.5	14.31	19.08	22.90	28.63	38.17	45.8	57.25	68.7	76.3	85.9	91.6	95.4	100.2
114.6	14.33	19.10	22.92	28.65	38.20	45.84	57.3	68.8	76.4	86.0	91.7	95.5	100.3
114.7	14.34	19.12	22.94	28.68	38.23	45.88	57.35	68.8	76.5	86.0	91.8	95.6	100.4
114.8	14.35	19.13	22.96	28.70	38.27	45.92	57.4	68.9	76.5	86.1	91.8	95.7	100.5
114.9	14.36	19.15	22.98	28.73	38.30	45.96	57.45	68.9	76.6	86.2	91.9	95.8	100.5
115.0	14.38	19.17	23.00	28.75	38.33	46	57.5	69.0	76.7	86.3	92.0	95.8	100.6

第十二章 刷表漆

Chapter XII
Surface Coating

用橘子油将生漆调配到合适
的浓稠度。

Mix the raw lacquer with
mandarin oil to proper consistency.

用发刷将漆均匀地布于琴面，一般顺序为先刷琴底，再刷琴面。

Use a hairbrush to evenly apply the lacquer on the qin, usually on the bottom first and then the surface.

采用横竖交替刷的方式将漆分布均匀。

The lacquer is evenly applied by brushing both horizontally and vertically, in an alternate manner.

蘸少许生漆刷琴的侧面，同时注意将刷正面时留下的生漆刷匀。

Stick some raw lacquer to brush the sides of the qin, while leveling off those left when brushing the qin surface and bottom.

用发刷均匀布漆，再将漆均匀刷开。

Use a hairbrush to evenly apply the lacquer, and then brush it evenly.

用横竖交替方式均匀涂刷。

Brush both horizontally and vertically in an alternate manner.

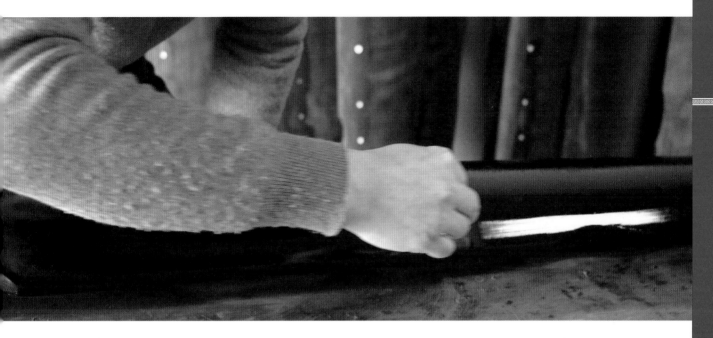

　　涂刷均匀后，用发刷轻轻在漆面进行长刷，拉平刷痕、小气泡等，然后静置流平一会儿放入阴房等待干燥。

　　After brushing evenly, use a hairbrush to slightly brush the lacquer surface to flatten the brushmarks, small bubbles, etc., then let the qin stand for a while and put it into a cool room to wait for the lacquer to dry

生漆是在自然生长的漆树上采集的天然树脂，直接使用，渗透性比较好。熟漆是生漆经过晾晒和熬制而成的，亮度比较高。一般来说靠木漆、灰胎都使用生漆，表漆多使用熟漆。

　　第一遍表漆也多使用生漆，用稀释剂（橘子油、松节油等）调和，生漆浓度要低一些，便于灰胎的渗透和吸收，增强灰胎的坚固性和表漆、灰胎的结合度。第一遍表漆干燥以后用1000目水砂纸进行通体打磨，第二遍开始刷熟漆，刷漆的时候先使漆均匀地分布在琴面，然后由漆多的地方向漆少的地方刷，再横竖交替逐步刷匀，最后用发刷轻轻在漆面进行长刷，拉平刷痕、小气泡等。用漆量一次不宜太多、太厚，以免出现流漆和起皱。侧面施工时注意正面的流漆，并注意不能再流漆到已刷好的正面，如再有流漆，用漆刷轻轻收回、刷平。每一遍刷漆结束后需静置半小时左右，等待自然流平再放入阴室，干燥以后再次使用更细的水砂纸通体打磨。

　　表漆一般刷三遍以上，讲究薄漆多刷，原则上来说刮最后一遍灰胎层后达到平整，刷第一、二遍表漆后达到光洁，刷三遍表漆后达到一定的亮度，这就是髹漆界流传的"一平，二光，三亮"的口诀。古琴表漆面不建议太亮，以亚光为好，这样经过长时间使用后达到的包浆效果才是最完美的。

Raw lacquer is a natural resin collected from naturally growing lacquer trees, which, with good permeability, can be used directly. Processed lacquer is obtained by drying and boiling raw lacquer, which has high brightness. Generally speaking, raw lacquer is used for wood lacquer and lacquer cement, and processed lacquer for surface lacquer.

Raw lacquer is also used for first-layer surface lacquer, which is blended with diluent (mandarin oil, turpentine, etc.). The concentration of raw lacquer should be lower to make it easily absorbed by the lacquer cement, so as to enhance the firmness of lacquer cement and the combination of surface lacquer and lacquer cement. After the first-layer surface lacquer is drying, polish the whole qin body with 1000 mesh water sandpaper. Starting from the second time, processed lacquer should be applied. First, apply the lacquer on the qin surface, then brush it horizontally and vertically, in an alternate manner, step by step. Finally, use a hairbrush to slightly brush the surface to flatten brush marks, small bubbles, etc. The amount of lacquer used at one time should not be too much or too little, so as to avoid lacquer flowing and wrinkling. Be careful not to touch the flow lacquer on the front when coating the sides, and prevent the lacquer from flowing to the brushed front. If there is additional flow lacquer, brush it gently with a paintbrush. Let the qin stand for about half an hour after each coating for the lacquer to level off, then put it into a cool room. After drying, polish the whole qin body again use finer water sandpaper.

Generally, the surface lacquer should be coated more than three times, with thin lacquer applied each time. In principle, at the last time, the lacquer cement layer will be flat; at the first and second times for coating, the surface lacquer will be smooth, and at the third time it will get a certain brightness - a pithy formula of "flat for the first time, smooth for the second and bright for the third" circulated in the qin decorating world. The surface lacquer of the Guqin should not be too bright, matte is preferred. The coating effect will be perfect after long-term use.

第十三章 磨表漆

Chapter XIII
Surface Paint
Polishing

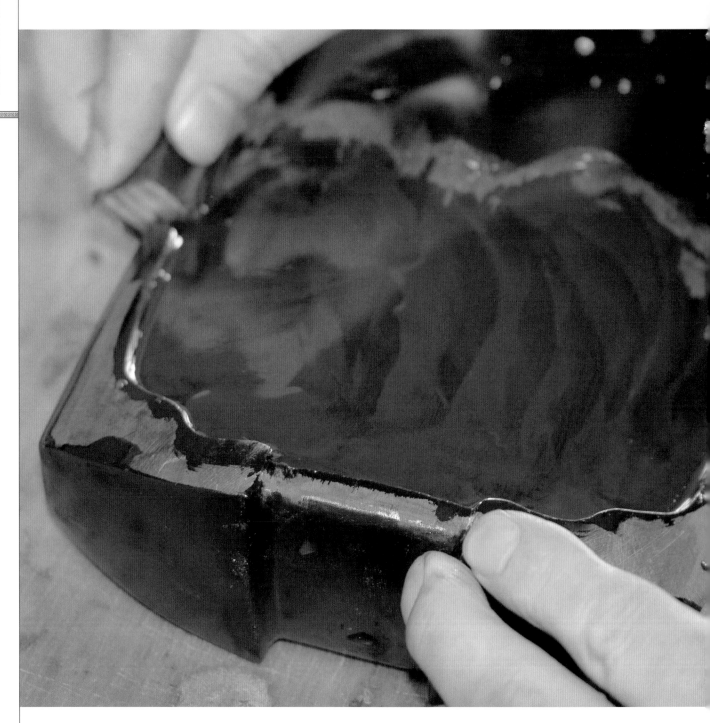

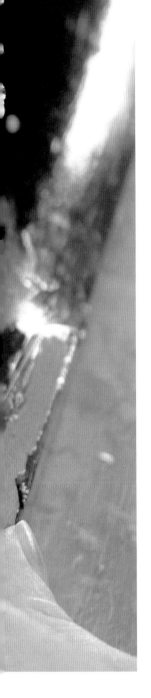

　　刷好的表漆干燥以后，如果漆面留有刷痕、尘粒等，可选用 800 目水砂纸垫小木方打磨。如果漆面较光洁，无刷痕、尘粒，选用 1000 目水砂纸垫橡皮进行打磨。

If there are brush marks, dust particles, etc. on the paint surface after drying, polish it with 800-mesh waterproof abrasive paper wrapping a small piece of wood. If the paint surface is smooth and free of brush marks and dust particles, polish it with 1,000-mesh waterproof abrasive paper wrapping a rubber.

　　打磨的时候，砂纸沾水采用椭圆形转圈的方式进行，一圈压一圈，均匀推进，并用湿毛巾擦拭，观察是否磨到位，是否存在用力过猛磨破等情况。

　　When polishing, dip the sandpaper in water and push forward evenly in oval circles, one by one, and wipe with a wet towel. Then observe whether it is well polished, or worn out due to overexerting.

打磨侧面和琴头、琴尾的时候必须轻柔而反复地进行，避免磨破表漆。

The sides, head and tail of the qin must be polished gently and repeatedly to avoid wearing out the surface paint.

护轸部位的打磨除
了垫小木方、橡皮以外，
也可以用手指直接裹砂
纸沿着弧度进行打磨。

When polishing the
Huzhen, in addition to
wood pieces and rubber,
you can also wrap your
fingers with the sandpaper
directly for polishing.

"斫琴亦斫心，磨漆即修行"，磨漆是一件十分考验人耐心的工作，需要坐得住，静得下心，沉得住气，一点一点地进行。

磨漆是否省力，刷漆工作十分重要，刷好的漆阴干以后不能有刷痕、起皱、流漆等情况，如果出现了这些问题，用 600 至 800 目水砂纸垫小木方整体打磨，打磨到刷痕、起皱、流漆等消失，整体平整为止。如果刷好的漆阴干以后表面光洁，无刷痕、起皱等情况，选用 1000 目水砂纸垫橡皮进行均匀打磨即可，打磨完再上第二遍漆，阴干以后再用 1500 至 2000 目砂纸打磨，第三遍砂纸打磨需要更细，逐步达到"平、光、亮"的效果。

打磨的时候选择光照良好的地方，用力必须均匀，不可过重，砂纸沾水后采用椭圆形转圈的方式，一圈压一圈，均匀推进，同时用湿毛巾擦拭，观察是否打磨到位，是否存在用力过猛、多次重复导致磨破等情况。未打磨的漆面反光，打磨以后呈亚光，所以在良好的光照下，是否打磨到位一目了然，如有反光点、面，则需继续打磨。琴头、尾和边角处最容易磨破，导致灰胎外露，必须更加耐心轻柔地反复打磨。

"The very process of Guqin making is one of self-cultivation, and grinding the paint surface is touching with your spiritual side." Paint surface polishing demands patience, in other words, you need to sit still, calm down, and do it bit by bit.

Whether paint polishing is labor-saving or not is up to the painting process. After the paint is dried in the shade, there should be no brush marks, wrinkles, paint flow, etc. If they occur, use 600-800-mesh waterproof abrasive paper wrapping a small piece of wood for overall polishing until the brush marks, wrinkles, paint flow, etc. are leveled off and the surface is flat. If the fresh paint is smooth and free of brush marks, wrinkles, etc. after drying in the shade, polish evenly with 1,000-mesh waterproof abrasive paper wrapping a rubber. Then paint the second time, and use 1,500-2,000-mesh sandpaper for second-time polishing after drying in the shade. When polishing for the third time, the sandpaper needs to be thinner to achieve the effect of "flat, smooth, and bright".

When polishing, choose a well-lit place and the strength applied should be equal. Do not exert too much force. Dip the sandpaper in water and push forward evenly in oval circles, one by one, and wipe with a wet towel. Then observe whether it is well polished, or worn out due to overexerting or repeated polishing. Unpolished paint reflects light, while the polished is matte. Therefore, it is clear at a glance whether the paint is polished or not in place with good illumination. If there are reflective spots and surfaces, it is necessary to continue polishing. The head, tail and corners of the qin tend to be worn out easily, resulting in the exposure of the lacquer cement. So these parts must be polished patiently, gently and repeatedly.

第十四章 擦表漆

Chapter XIV
Surface Paint
Wiping

用新棉纱清理干净琴面的杂质、灰尘等。

Clean up the impurities and dust on the qin surface
with a piece of new cotton yarn.

用棉纱蘸少量生漆，将其均匀分布在琴面上。

Use the cotton yarn to dip a small amount of raw lacquer and evenly apply it on the qin surface.

Smear the raw lacquer evenly.

将生漆涂抹均匀。

Smear the raw lacquer evenly.

注意琴头、尾和护轸都需
要涂抹到位。

Pay special attention to the
qin head, tail and Huzhen parts.

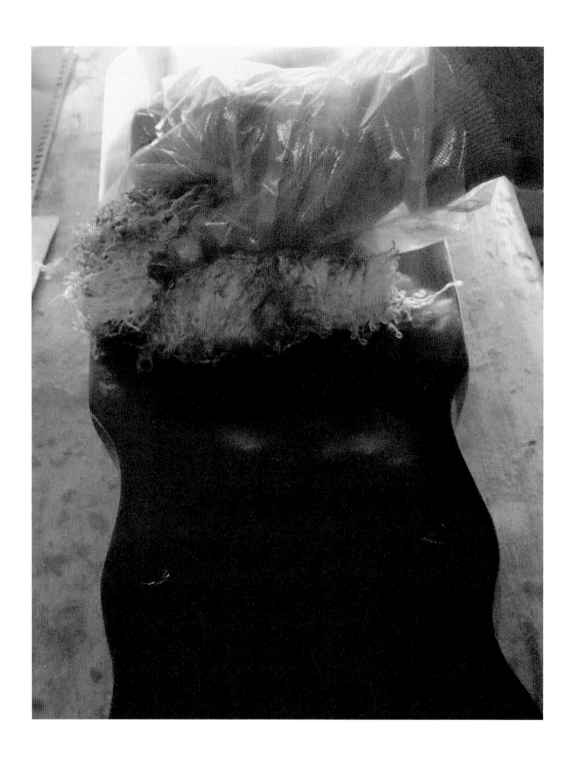

然后用大块棉纱收掉多余的生漆。

Then use a large piece of cotton yarn to absorb excess raw lacquer.

换干净棉纱进一步擦拭，收掉多余生漆。

Change another piece of clean cotton yarn to absorb excess raw lacquer.

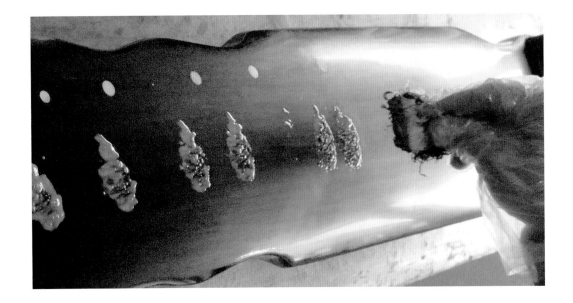

用同样的方法擦拭另一面。

Polish the other side in the same way.

蘸漆涂抹的棉纱球小一点。

Make sure that the cotton yarn used to dip raw lacquer is small in size.

收漆的棉纱团大一点，回收多余生漆的时候，逐步往后换干净棉纱。

And the cotton yarn ball used for raw lacquer absorption is much larger. When absorbing the surplus raw lacquer, change these cotton yarn balls in turn from left to right.

擦漆是古琴髹漆施工的最后一步，主要为了达到光亮美观的效果。一般擦漆使用花生油或煤油进行稀释，它们能够增加漆液的润滑度，便于施工，花生油还可以提高亮度效果，煤油可以提高渗透性。

擦漆原则是多遍、少量、涂抹均匀、收干净，不能在漆面留下眼睛能看见的漆液，一般一张琴擦漆六遍以上效果才会逐渐显现出来。擦漆工作其实可以参考生活中擦皮鞋的经验，二者有许多原理相同，不同之处就是擦漆需要反复多遍进行。

除了擦漆法，最后一遍的漆面处理还可以使用推光法，就是将经过砂纸水磨后的漆面，少量而均匀地涂抹上油脂，然后使用极细的瓦灰或者面粉进行推光，用于反复进行，直到光亮为止。

Lacquer polish is the last step in Guqin painting which can make the qin look more bright and appealing. Generally, the lacquer is diluted by peanut oil or kerosene to increase the lubricity of the paint and facilitate construction. Peanut oil can also improve its brightness and kerosene permeability.

The rules are to apply many times, in a small amount and an even and clean manner, leaving no paint liquid visible to the eyes on the qin surface. The effect can be seen only after six or more polishes. In fact, lacquer polishing is much like shoe polishing in life, the two share similar principles, and the only difference is that lacquer polish should be repeated many times.

In addition to the lacquer polish method, the push light method can also be used for last-time paint surface treatment. That is, apply a small amount of grease on the qin surface sanded down by waterproof abrasive paper, sprinkle a layer of fine tile ash or powder and push repeatedly by hand until the surface turns bright.

第十五章　上琴弦

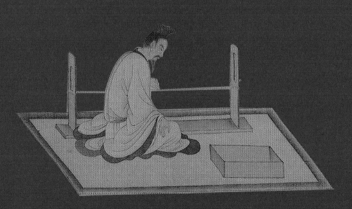

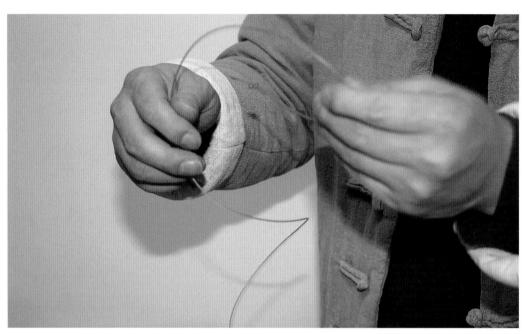

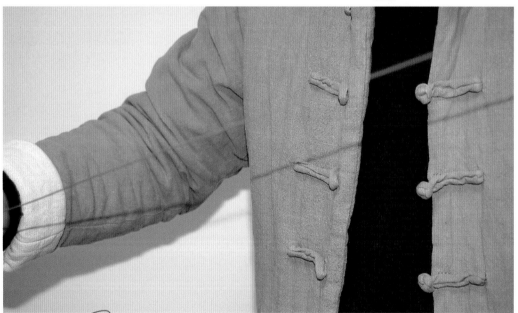

　　琴弦在打蝇头和上弦的时候一定注意避免在弦路位置出现打结、对折等情况，以免上好弦以后出现煞音。

When stringing and tying the butterfly knot, be really careful not to tie any knot on or fold the string path so as to avoid brakes.

　　将打好蝇头的琴弦穿到绒扣中，调整到岳山前（靠一徽位置）三分之一处，可根据自己上弦松紧习惯提前预留好调音的位置。

　　Thread the strings with butterfly knots into the velvet buckle and adjust them to one-third of the bridge (near the first emblem). You can reserve the tuning position in advance according to your requirements on tightness.

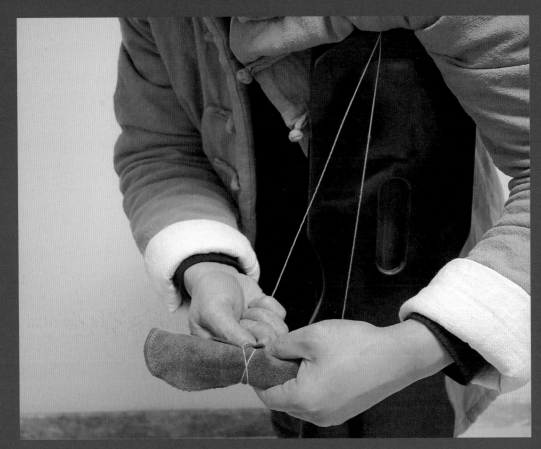

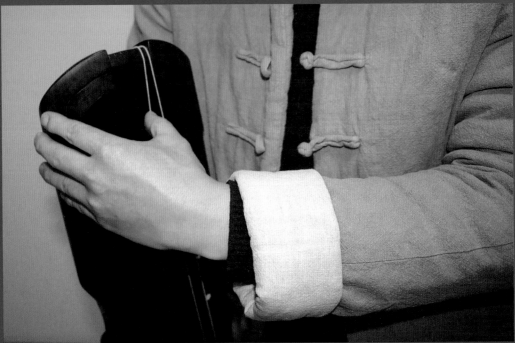

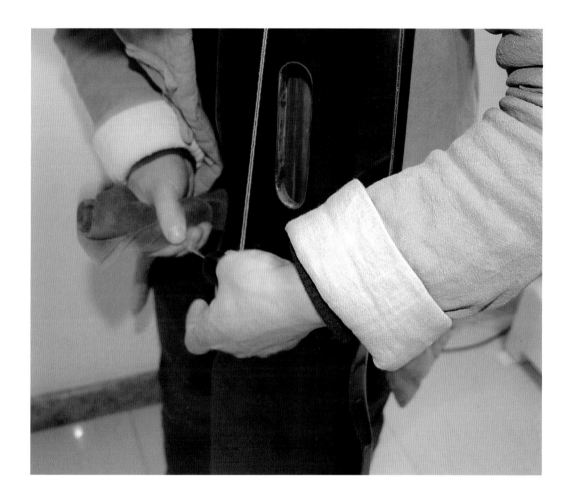

　　将琴弦末端缠到拉弦器上，再将琴弦绕雁足半圈，用适当力气拉紧琴弦，一边拉一边用另一只手拨弦，待琴弦拉紧到相应音高以后（音高可用传统定音哨或者另一张已调整为标准音的琴进行参考），可适当高一点，再将弦缠绕于雁足上，缠绕的时候注意不要让琴弦出现松滑。

　　Wrap the end of the strings around the stringer, and then wind the strings around the wild goose feet for half a circle. Tighten the strings with an appropriate strength, and pluck the strings with the other hand while tightening. After the strings are tightened to a position with a certain pitch (by reference to the traditional tuning whistle or another qin that has been adjusted to the standard tone) - a higher pitch is allowable - and then wrap the strings on the wild goose feet. Be careful not to let the strings slip.

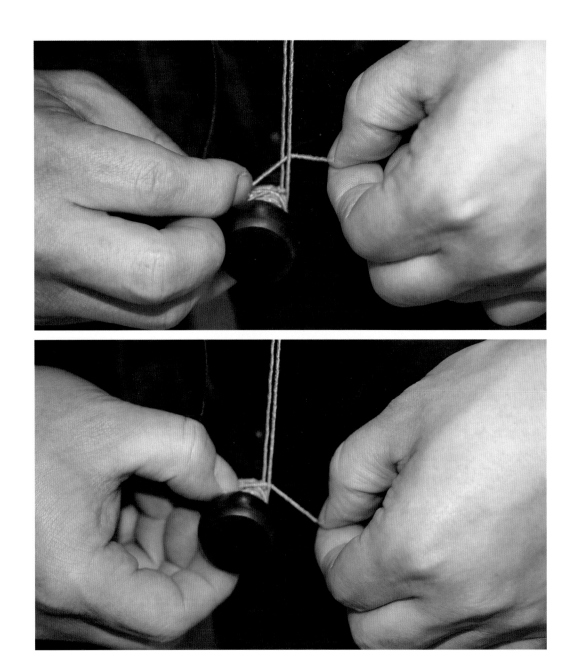

将缠绕于雁足的琴弦末端卡住，防止松脱，前几根弦的末端可在下一根弦缠绕的时候压于下端，最后一根弦可多卡几次或者打结防止松脱。

Stick the end of the strings wrapped around the wild goose feet to prevent loosening. The end of the former strings can be covered by the next string wound. The last string can be stuck or knotted several times to prevent loosening.

　　注意上好后每根弦的蝇头位置必须在岳山前三分之一和二分之一处之间。

　　Note that the position of each string's butterfly knot must be between one-third and one-half of the bridge.

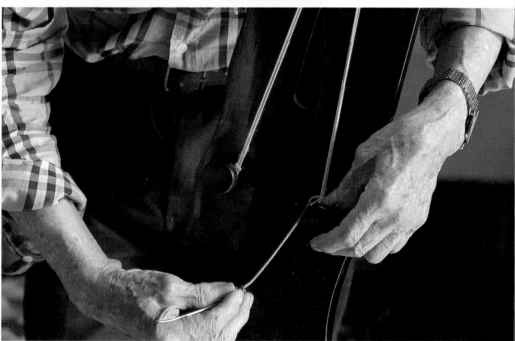

琴弦最后打结和卡的时候可以使用穿绒扣器上的小钩，这样更方便。

When the strings are finally knotted and stuck, the small hook on the velvet fastener is much helpful.

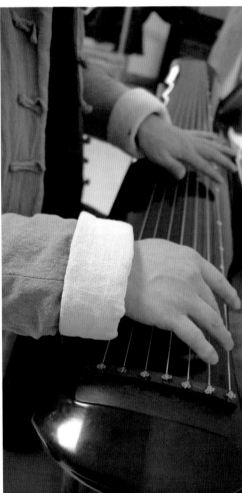

用定音哨对各弦进行初步调音，然后采用泛音调音法调好各弦音高。

Each string is preliminarily tuned with a tunning whistle, and then the pitch of each string is tuned with the overtone tuning method.

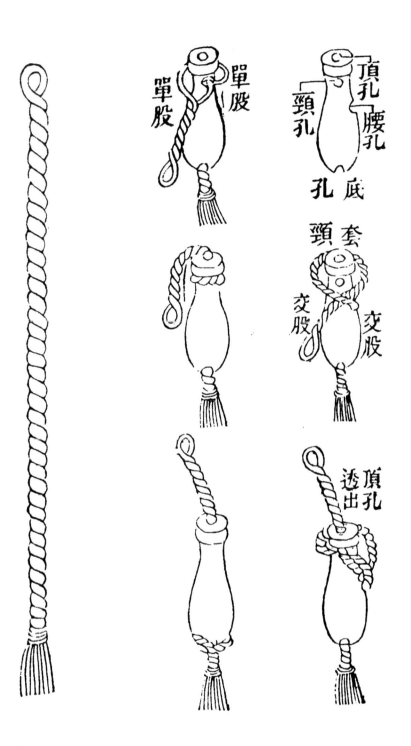

清代《琴学入门》绒扣穿轸法示意图

Schematic Diagram for Stringing *with a Velvet Buckle in An Introduction to Qin Studies of Qing Dynasty*

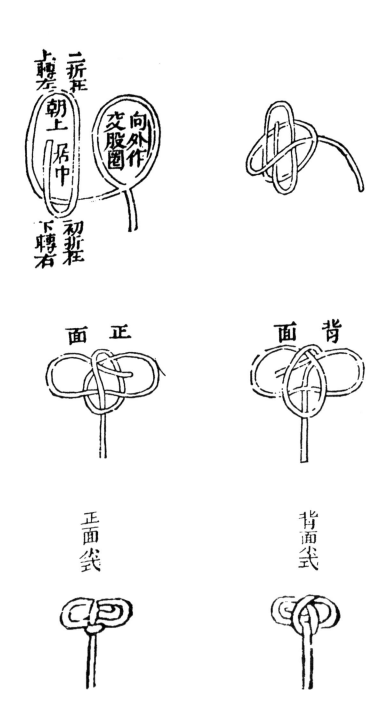

清代《琴学入门》结蝇头法示意图

Schematic Diagram for Butterfly Knot Tying in *An Introduction to Qin Studies* of Qing Dynasty

"伏羲削桐为琴，绳丝为弦"，琴从开始出现就以蚕丝为弦，20 世纪 70 年代研发了钢丝尼龙弦，90 年代又研发了纯尼龙弦，这三种材质的琴弦各有特点。

传统丝弦声音古朴，散、泛、按三音更为和谐，但因用蚕丝绞合而成的琴弦表面粗糙，手感较差，容易断裂，需精心护理。老先生在上新弦前会选用细砂纸粘胶水（一种用鱼鳔、白芨等熬制调配而成的琴弦专用胶）进行打磨，让丝弦弹奏时手感更好一些，减少过手杂音。

钢丝尼龙弦声音清亮，泛音有所损失，不易断，无须特殊护理，现已普遍使用。钢丝尼龙弦受钢丝质量影响比较大，过去质量较好的为德国"蓝牌"钢丝。

纯尼龙弦音色介于传统丝弦和钢丝尼龙弦之间，有时略显迟钝，灵敏度稍微差一点，其拉力良好，不易断裂。

It was recorded that "Fuxi cut sycamore as a qin and strung it with silken cords", hence people first used silk to string qins. Afterwards, steel-nylon strings were developed in the 1970s and pure nylon strings in the 1990s. Each of the three sorts of strings has its distinct features.

The traditional silk strings are simple and unsophisticated in sound, and have more harmonious scattering, over, and pressing tones. Yet stranded by silk fibers, they have a rough surface and poor hand feeling, and can easily break, thus requiring meticulous care. Master makers, when stringing a qin, will polish the silk strings with fine sandpaper pasted with glue (a special glue for strings made by boiling swimming bladder, bletilla striata, etc.) to make the strings feel better and reduce noise caused by hand when playing.

Steel-nylon strings, though with a clear sound, are inferior to the other two in overtones but not easy to break, and no special care is needed. Such strings are now widely used. Steel-nylon strings are greatly affected by the quality of steel wire. In the past, Germany's "ROSLAU" steel wire got good word of mouth due to its high quality.

The timbre of pure nylon strings is intermediate between traditional silk strings and steel-nylon strings. Slightly less sensitive, though, pure nylon strings have good tensile strength and are not easy to break.